SORPTION ISOTHERMS and WATER ACTIVITY of FOOD MATERIALS

SORPTION ISOTHERMS
and
WATER ACTIVITY
of
FOOD MATERIALS

Compiled by

W. WOLF, W. E. L. SPIESS and G. JUNG

ELSEVIER

NEW YORK . AMSTERDAM . OXFORD

A bibliography related to the use, application, measurement and theory of equilibrium water conditions in food materials prepared within the framework of COST-90 and 90 bis Projects subproject 'Water activity' in cooperation with H. Bizot, M. C. Gough, Y. Motarjemi, M. Rüegg, H. Weisser and with the assistance of the Project Leader Ronald Jowitt.

ISBN 0-444-00997-3

Published by Science and Technology Publishers Ltd, 1985
33 Woodlands Avenue, Hornchurch, Essex RM11 2QT, England

Sole Distributors in North America:
Elsevier Science Publishing Co. Inc.,
52 Vanderbilt Avenue, New York, New York 10017

Printed in Great Britain by Galliard (Printers) Ltd, Great Yarmouth

Contents

* The sub-chapters 3.1, 3.2, 3.3, 3.4 and 3.5 correspond to the code numbers 1, 2, 3, 4 and 5 always given in the last line of the cited references.

1 Introduction

General remarks

Water is the constituent present in the highest concentration in fresh foods; it influences the palatability, digestibility, physical structure and technical handling ability of the food material considerably. Most important, however, practically all deteriorative processes that take place in foodstuffs are influenced in one way or another by the concentration and mobility of water. As a rule of thumb it can be said that, independent of the composition of food, the mechanisms of spoilage at high water concentrations are in microbiological nature (growth of bacteria, yeasts and moulds) and of enzymic or nonenzymic origin; the mechanisms of spoilage at low water concentrations are mainly autoxidative reactions. The intensity and rate of the various deteriorative processes are different at different water concentrations. In general food material is more stable at low than at high water concentrations.

Water activity

The potential of water to take part in deteriorative processes in a food product is characterized by the activity of the water in the food which is, according to the generalized Raoult's law, the ratio between the water vapor pressure of the product at a given temperature and the saturation pressure of pure water at the same temperature. A general relationship between the water activity and the kind of deterioration cannot be established, but water activity ranges can be defined where certain types of deteriorative reactions are dominant (Labuza, T. P., 1137).

The water activity of any material depends on the chemical composition of the product, the state of aggregation of its constituents, the water content, and the temperature of the material.

Sorption isotherm

The relationship between the water content of a material and its corresponding water activity at a given temperature is expressed by sorption isotherms results from their usefulness in various sectors of food technology; knowing the water-vapor sorption isotherms of sorption isotherms results from their usefulness in various sectors of food technology; knowing the water-vapor sorption isotherms of food materials, it is possible to calculate sorption and desorption enthalpies, to determine "bound water", to check the crystalline structure of products, to facilitate the calculation and operation of drying, mixing and packaging processes, to evaluate the physical, chemical and microbiological stability of the product and to predict its shelf life.

The sorption isotherms of most food materials are in sigmoid shape. Since the water sorption behavior of food is determined by its chemical composition and the physicochemical state of the food constituents, the sorption isotherms of different food products differ considerably from each other: Starchy products including cereals absorb more water at low water activities than protein-rich materials like meat. Sugars, in general, represent a group of materials that exhibit two types of sorption behavior. In the amorphous state they are more hygroscopic, than in the crystalline state. Similar products may show differences in the shape of their sorption isotherm if they are of different origin or if they have been treated differently.

In general, the water binding ability of a product is reduced by any treatment which includes a thermal process step since the number of active sites capable of binding water is usually reduced by such processes as heating and freezing, as it is by processes such as desalting and changes in pH.

Although the phenomena of water sorption by foodstuffs have attracted the interest of numerous research workers, a consistent thermodynamic theory for these phenomena is still lacking. The reason for the failure of most of the theoretical models is that foodstuffs have rather complicated chemical compositions and also very complex physical structures.

The sorption of water by foodstuffs can be classified into three main categories: (a) sorption without structural changes of the sorbent, e.g., surface sorption in the case of crystalline sugar; (b) sorption accompanied by structural changes of the sorbent, e.g. milk, egg white; and (c) sorption formation of a solution, e.g. sugar solutions.

Theoretical approaches to, or leading to, the sorption behavior of foodstuffs have been reviewed by Van den Berg, C. and Bruin, S. (2045). The most popular theory is the BET approach (Brunauer, S.; Emmet, P. H.; Teller, E., 289). Despite the fact that most of the assumptions on which the theory is based are not valid for food materials, its use is of some practical value. The so called monolayer value, for example—which can be calculated by means of the BET equation—characterizes for a great number of food products the water activity range in which the food is most stable.

If adsorption and desorption data are plotted together, the adsorption curve nearly always shows lower values for the water content than the desorption curve at the same water activity values. This hysteresis phenomenon is another fact which demonstrates the complex binding of water in food materials. To explain the hysteresis effect, again several theories have been proposed. With porous material, hysteresis can be due to a change in the contact angle between adsorption and desorption when capillary condensation in pores or evaporation from pores occurs. The basis for these theories is the Kelvin equation representing the vapor pressure depression in pores. For nonporous materials, hysteresis can be explained by energy-consuming processes in the interior of a product such as swelling, relaxation processes, etc. For both porous and nonporous materials, hysteresis may be explained in terms of phase changes.

The precise determination of the relationship between the water activity of a foodstuff and its equilibrium moisture content is a simple procedure which requires, nevertheless, experience and experimental skill.

A comprehensive review of existing measuring methods is given by S. Gál (660).

In the last four years a group of experts working together in the COST-90 Project has developed a reference material (microcrystalline cellulose, MCC) and a reference method for measuring water-vapor sorption isotherms and conducted a collaborative study to determine the precision (repeatability and reproducibility) with which the sorption isotherm of the reference material may be determined by the reference method (Wolf, W. et al., 2160; Spiess, W. E. L., Wolf, W., 1911). Sorption isotherms are gaining in importance as the number of recommendations, official regulations and product specifications which use water activity as an evaluation criterion is growing continuously.

Sorption bibliography

The present bibliography attempts to make the literature data and information accessible to the interested industry and research institutes. The extensive bibliography on questions of water sorption by foods and related products is based on a collection of water vapor sorption isotherms (Wolf, W. *et al.*, 2158) which was supplemented continuously in the following years. Finally the collection was extended by contributions made by members of the Working Group on Water Activity of COST-90. The laboratories participating in the project had, in compliance with one of the project goals, compiled information concerning water activity of foods, in order to make these data available to industry in the participating COST member countries. The collection now comprises 2,201 references including about 900 papers with information on equilibrium moisture contents of foods in defined environments.

The aim has been to present as complete a list as possible of publications dealing with water vapor sorption, and to procure the papers, as far as possible, for evaluation. Any papers not directly concerned with the subject were included only if they contained fundamental contributions or information on pertinent literature.

The original intention to revise and supplement the first data collection by Wolf *et al.* so that the latest sorption data could be included was deferred because the material available had become so voluminous and because uncritical publication of it seemed inadvisable.

Although other bibliographies have been published recently (Heiß, R., 844; Kneule, F., 1094; Iglesias, H. A. and Chirife, J., 2201) the present collection contains the greatest number of references so far.

The present collection, contains in Chapter 2, a list of all publications arranged alphabetically according to the names of the first author. The list is numbered consecutively, and each reference is supplemented by a code number indicating the subject of the paper.

In Chapter 3 papers are arranged according to subjects. The data on equilibrium moisture contents of food in defined environments (Chapter 3.5) are also grouped according to products and product groups, respectively. (Chapter 3.5.2).

Reprints of most of the papers included are held at Bundes-forschungsanstalt für Ernährung (the Federal Research Centre for Nutrition). Copies of individual papers are sent upon request in accordance with international copyright agreements.

2. Bibliography

Arranged alphabetically according to first authors; numbered consecutively

1 Abadie, P.; Charbonniere, R.; Gidel, A.; Girard, P.; Guilbot,A.
   L'eau dans la cristallisation du maltose et du glucose et
   états de l'eau de sorption de l'amidon d'après les spectres
   d'absorption en radiofréquences.
   Journal de Chimie Physique 50 (7-8) C46-C52; 1953
   2
2 Abramochkin, A.I.
   Utilization of sorption isotherms for the determination of
   optimum moisture content of spray-dried caseinate. (orig. russ.)
   Trudy, Vsesoyuznyi Nauchno-issledovatel'skii Institut Molochnoi
   Promyshlennosti 49 (-) 4; 1979
   3,5
3 Abrams, D.; Prausnitz, J.
   Statistical thermodynamics of liquid mixtures: a new
   expression for the excess Gibbs energy of partly or completely
   mixible systems.
   AICHE Journal 21 (1) 116-128; 1975
   2
4 Aceto, N.C.; Sinnamon, H.J.; Schoppet, E.F.; Craig, J.C.;
   Eskew, R.K.
   Moisture content and flavor stability of batch vacuum
   foam-dried whole milk.
   Journal of Dairy Science 48 (5) 544-547; 1965
   4,5
5 Acheson, D.T.
   Vapor pressures of saturated aqueous salt solutions.
   In: Wexler, A.; Wildhack, W.A. (Edts.). Humidity and Moisture
   Measurement and Control in Science and Industry. Fundamentals
   and Standards. Vol. 3, pp 521-530.
   Reinhold Publ. Corporation, New York; 1965
   3
6 Acker, L.
   Die Lagerfähigkeit lufttrockener Lebensmittel als Funktion der
   Feuchtigkeit.
   Die Nahrung 2 (-) 1045-1061; 1958
   1,4,5
7 Acker, L.
   Enzymatische Reaktionen in festen Stoffen -
   ein lebensmittelchemisches Problem.
   Angewandte Chemie 74 (15) 592; 1962
   1
8 Acker, L.
   Enzymic reactions in foods of low moisture content.
   Advances in Food Research 11 (-) 263-330; 1962
   4
9 Acker, L.
   Enzyme activity at low water contents
   In: Leitch,M.J.; Rhodes, D.N. (Edts.).
   Recent Advances in Food Science. Vol. 3, pp 239-247; 1963
   4,5
10 Acker, L.
   Über die Beziehung zwischen Enzymaktivität und Wasseraktivität
   in trockenen Lebensmitteln.
   Die Nahrung 12 (5) 557-564; 1968
   1,4,5
11 Acker, L.
   Enzymatische Veränderungen in wasserarmen Lebensmitteln.
   In: Enzyme und Ernährung; Dr. Dietrich Steinkopff-Verlag,
   Darmstadt; 1969
   4

12 Acker, L.
    Microbiological and biochemical aspects at low water
    activities in dehydrated foods.
    Dechema-Monographien 63 (1125-1143) 203-218; 1969
    1,4,5

13 Acker, L.
    Water activity and enzyme activity.
    Food Technology 23 (10) 1257-1270; 1969
    1,4,5

14 Acker, L.
    Nährmittel als reaktive Systeme im Gleichgewicht mit ihrer
    Umgebung.
    Getreide, Mehl und Brot 33 (6) 159-163; 1979
    1,4,5

15 Acker, L.; Beutler, H.
    Die enzymatische Phytinspaltung in geschrotetem Getreide in
    Abhängigkeit von der relativen Luftfeuchtigkeit.
    Zeitschrift für Ernährungswissenschaft 3 (-) 1-5; 1963
    4

16 Acker, L.; Beutler, H.O.
    Über die Tätigkeit der Lipasen in Getreidemahlprodukten in
    Abhängigkeit von der relativen Luftfeuchtigkeit.
    Getreide und Mehl 15 (1) 4-7; 1965
    4,5

17 Acker, L.; Beutler, H.O.
    Über die enzymatische Fettspaltung in wasserarmen
    Lebensmitteln.
    Fette, Seifen, Anstrichmittel 67 (6) 430-433; 1965
    4,5

18 Acker, L.; Huber, L.
    Über das Verhalten der Polyphenoloxydase im wasserarmen
    Milieu.
    Lebensmittel-Wissenschaft und Technologie 2 (-) 82-85; 1969
    4

19 Acker, L.; Huber, L.
    Über das Verhalten der Glucoseoxidase im wasserarmen Milieu.
    Lebensmittel-Wissenschaft und Technologie 3 (-) 33-36; 1970
    4,5

20 Acker, L.; Kaiser, H.
    Über den Einfluss der Feuchtigkeit auf den Ablauf
    enzymatischer Reaktionen in wasserarmen Lebensmitteln.
    Zeitschrift für Lebensmittel-Untersuchung und -Forschung
    110 (5) 349-356; 1959
    4

21 Acker,L.;Kaiser,H.
    Über den Einfluss der Feuchtigkeit auf den Ablauf
    enzymatischer Reaktionen in wasserarmen Lebensmitteln.
    III.Mitteilung:Das Verhalten der Phosphatase in
    Modellgemischen.
    Zeitschrift für Lebensmittel-Untersuchung und -Forschung
    115 (-) 201-210; 1961
    4,5

22 Acker, L.; Lueck, E.
    Über den Einfluss der Feuchtigkeit auf den Ablauf
    enzymatischer Reaktionen in wasserarmen Lebensmitteln.
    Zeitschrift für Lebensmittel-Untersuchung und -Forschung
    108 (3) 256-269; 1958
    4,5

23 Acker, L.; Wiese, R.
Über das Verhalten der Lipase in wasserarmen Systemen.
2. Mitteilung. Enzymatische Lipolyse im Bereich extrem
niedriger Wasseraktivität.
Zeitschrift für Lebensmittel-Untersuchung und -Forschung
150 (4) 205-211; 1972
4

24 Acker, L.; Wiese, R.
Verhalten der Lipase im wasserarmen Milieu.
1. Einfluß des Aggregatzustandes des Substrates auf die
enzymatische Lipolyse.
Lebensmittel-Wissenschaft und -Technologie 5 (5) 181-184; 1972
4

25 Acott, K.M.; Labuza, T.P.
Microbial growth response to water sorption preparation.
Journal of Food Technology 10 (6) 603-611; 1975
4

26 Acott, K.M.; Labuza, T.P.
Inhibition of Aspergillus niger in an intermediate moisture
food system.
Journal of Food Science 40 (1) 137-139; 1975
4

27 Acott, K.M.; Sloan, A.E.; Labuza, T.P.
Evaluation of antimicrobial agents in a microbial challenge
study for an intermediate moisture dog food.
Journal of Food Science 41 (3) 541-546; 1976
4

28 Adam, S.; Blankenhorn, R.; Diehl, J.F.
Effect of water activity upon electronradiolysis of dextran.
Starch/Stärke 31 (12) 423-428 ; 1979
4,5

29 Adams, G.H.; Ordal, Z.J.
Effects of thermal stress and reduced water activity on
conidia of Aspergillus parasiticus.
Journal of Food Science 41 (3) 547-550; 1976
4

30 Adams, J.R.; Merz, A.R.
Hygroscopicity of fertilizer materials and mixtures.
Industrial and Engineering Chemistry 21 (4) 305-307; 1929
3

31 Adamson, A.
Adsorption of gases and vapors on solids; the surface area of
solids.
In: Physical Chemistry of Surfaces. Interscience Publ.,
New York; 1960
2

32 Adamson, A.
Physical Chemistry of Surface.
3.edition, John Wiley and Sons, New York; 1976
2

33 Adrian, J.
La réaction de Maillard vue sous l'angle nutrionnel.
Industries Alimentaires et Agricoles 89 (9-10) 1281-1289; 1972
4

34 Agrawal, K.K.
Drying characteristics of paddy with forced ventilation.
M.S. Thesis, Indian Institute of Technology, Kharagpur, India;
1967
3,5

35 Agrawal, K.K.; Clary, B.L.; Nelson, G.L.
    Investigation into the theories of desorption isotherms for
    rough rice and peanuts.
    Journal of Food Science 36 (6) 919-924; 1971
    2,5
36 Aguirre, J.M.; Travaglini, D.A.; Cabral, A.G.; Travaglini,
    M.M.E.; Silveira, E.T.F.; Sales, A.M.; Figueiredo, I.B. de;
    Ferreira, V.L.P.
    Secagem e armazenamento do residuo resultante do processamento
    do extrato de soja.
    Boletim do Instituto de technologia de alimentos, Campinas
    18 (2) 227-243; 1981
    5
37 Ahlgrimm, H.J.
    Zur dielektrischen Bestimmung des Feuchtigkeitsgehaltes von
    Lebensmitteln.
    Zeitschrift für Lebensmittel-Technologie und Verfahrenstechnik
    28 (-) 305-311; 1977
    3
38 Alam, A.; Shove, G.C.
    Hygroscopicity and thermal properties of soybeans.
    Transactions of the American Society of Agricultural Engineers
    16 (4) 707-709; 1973
    2,5
39 Alcaraz, C. de; Martin Martin, A; Pereda Marin, J.
    Metodo manometrico para medida de humedades de equilibrio.
    Grasas y Aceites 28 (6) 403-407; 1977
    2,3,5
40 Alderton, G.; Snell, N.
    Chemical states of bacterial spores: heat resistance and its
    kinetics at intermediate water activity.
    Applied Microbiology 19 (4) 565-572; 1970
    4
41 Alekseeva, L.; Bondar, V.; Surkova, N.
    The critical and equilibrium moisture content of triticale.
    Mukomol'no-elevatornaya i Kombikormovaya Promyshlennost'
    No. 2, 29; 1981
    5
42 Algie, J.E.
    Some properties of a sample of agar. Part II. The effect of
    absorbed water.
    Colloid and Polymer Science 257 (-) 117-120; 1979
    4
43 Allamand, A.J.; Hand, P.G.T.; Manning, J.E.
    The sorption of water by activated charcoals. Part V.
    Journal of Physical Chemistry 33 (-) 1694; 1929
    2,5
44 Almasi, E.
    Binding energy of bound water in foodstuffs.
    Acta Alimentaria 7 (3) 243-255; 1978
    2
45 Almasi, E.
    Dependence of the amount of bound water of foods on
    temperature.
    Acta Alimentaria 8 (1) 41-56; 1979
    2,3,5
46 Almeroth, K.
    Zur Bestimmung spezifischer Oberflächen mit dem Areameter.
    Chemie-Ingenieur-Technik 40 (23) 1181; 1968
    3

47 Altmann, R.L.; Benson, S.W.
The sorption of water vapor by native and denatured egg albumin.
Journal of Physical Chemistry 64 (-) 851-855; 1960
2,5

48 Alzamora, S.M.; Chirife, J.; Ferro-Fontan, C.
Effect of surface active agents on water activity of IM food solutions.
Journal of Food Science 46 (6) 1974-1975; 1981
2,4

49 Amberg, C.H.
Heats of adsorption of water vapor on bovine serum albumin.
Journal of the American Chemical Society 79 (-) 3980-3984; 1957
2,5

50 Amdur, J.E.; White, R.W.
Two pressure relative humidity standards.
In: Wexler, A.; Wildhack, W.A. (Edts.). Humidity and Moisture. Vol. 3, pp 445-454.
Reinhold Publ. Corporation, New York; 1965
3

51 Aminoff, C.; Vanninen, E.; Doty, T.E.
The occurence, manufacture and properties of xylitol.
In: Counsell, J.N. (Ed.). Xylitol. pp 1-9.
Applied Science Publ., London; 1978
5

52 Anagnostopoulos, G.D.
Water activity in biological systems - A dew-point method for its determination.
Journal of General Microbiology 77 (-) 233-235; 1973
1,4

53 Anagnostopoulos, G.D.
Water activity in food.
Nutrition and Food Science 59 (-) 6-7; 1979
1,4

54 Andales, S.C.; Pettibone, C.A.; Davis, D.C.
Influence of relative humidity and temperature on weight and sucrose losses of stored sugarbeets.
Transactions of the American Society of Agricultural Engineers 23 (2) 477-480; 1980
4

55 Anderson, C.B.; Witter, L.D.
Water binding capacity of 22 L-amino-acids from water activity 0.33 to 0.95.
Journal of Food Science 47 (6) 1952-1954; 1982
2,5

56 Anderson, C.B.; Witter, L.D.
Glutamine and proline accumulation by Staphylococcus aureus with reduction in water activity.
Applied and Environmental Microbiology 43 (6) 1501-1503; 1982
4

57 Anderson, J.A.; Alcock, A.W. (Edts.).
Storage of cereal grains and their products. Chapter 1
American Association of Cereal-Chemists, St. Paul, Minnesota; 1954
4,5

58 Anderson, J.A.; Babbitt, J.D.; Meredith, W.O.S.
The effect of temperature differential on the moisture content of wheat.
Canadian Journal of Research 21 C (-) 297-306; 1943
2

59 Anderson, J.H.; Parks, G.A.
    The electrical conductivity of silica gel in the presence of
    adsorbed water.
    The Journal of Physical Chemistry 72 (10) 3662-3668; 1968
    4,5
60 Anderson, R.B.
    Modifications of the Brunauer, Emmett and Teller equation.
    Journal of the American Chemical Society 68 (-) 686-691; 1946
    2
61 Anderson, R.B.; Hall, K.W.
    Modifications of the Brunauer, Emmett and Teller equation II.
    Journal of the American Chemical Society 70 (-) 1727-1734;
    1948
    2
62 Anderson, Y.
    The influence of moisture content on rheological properties of
    textured soy proteins.
    Proc. IV International Congress of Food Science and
    Technology 2 (-) 221-229; 1974
    4,5
63 Anet, E.L.F.J.; Reynolds, T.M.
    Reactions between amino acids and sugars in freeze dried
    apricots.
    Nature 177 (4519) 1082-1083; 1956
    4
64 Angeli, A.; Cirilli, G.; Fiorette, A.
    Die Bedeutung des freien Wassers für die Haltbarmachung von
    Backwaren. (Orig. ital.)
    Industria Alimentari 18 (3) 193-197; 1979
    2
65 Anisimova, L.V.; Yegorov, G.A.
    Hygroscopic properties of parts of the wheat kernel. (Orig.
    russ.)
    Pistschewaja Technol. - (2) 32-34; 1977
    5
66 Anker, C.A.; Geddes, W.F.; Bailey, C.H.
    A study of the net weight changes and moisture content of
    wheat flour at various relative humidities.
    Cereal Chemistry 19 (-) 128-150; 1942
    5
67 Antonov, A.; Losinskaya, N.V.; Grinberg, Y.; Dianova, V.T.;
    Tolstogusov.
    Phase equilibria in water-protein-polysaccharide systems.
    III Water-soy bean globulins polysaccharide systems.
    Colloid and Polymer Science 257 (-) 1159-1171; 1979
    2,4
68 Applebey, M.P.; Crawford, F.H.; Gordon, K.
    Vapor pressure of saturated solutions. Lithium chloride and
    lithium sulphate.
    Journal of the Chemical Society 11 (-) 1665; 1934
    3
69 Arai, S.; Yamashita, M.; Fujimaki, M.
    Protease action on proteins at low water concentration.
    In: Rockland, L.B..; Stewart, G.F. (Edts.). Water Activity:
    Influences on Foodquality. pp 489-510. Academic Press,
    New York; 1981
    4
70 Arkcoll, D.B.
    Preservation of leaf protein preparations by air drying.
    Journal of the Science of Food and Agriculture 20 (10)
    600-602; 1969
    5

71 Arkhangelskii, L.K.;Materova,E.A.
    Water vapor sorption by mixed forms of sulfonate cation
    exchangers with varying amounts of cross-linking.(Orig.russ.)
    Vestnik Leningradskogo Universiteta Servija Fiziki Khimii
    23 (10) 146-148; 1968
    2
72 Armstrong, A.A. jr.; Stannett, V.
    Temperature effects during the sorption and desorption of
    water vapor in high polymers.
    I. Fibres with particular reference to wool.
    Die Makromolekulare Chemie 90 (-) 145-160; 1966
    2
73 Armstrong, A.A. jr.; Wellons, J.P.; Stannett, V.
    Temperature effects during the sorption and desorption of
    water vapor in high polymers.
    Part II: Films with special reference to ethyl cellulose.
    Die Makromolekulare Chemie 95 (-) 78-91; 1966
    2
74 Arnell, J.C.; McDermot, H.L.
    Sorption hysteresis.
    Proc. of the 2. International Congress of Surface Activity.
    Vol. 2, Academic Press, New York; 1957
    2
75 Arya, S.S.; Natesan, V.; Pavihar, D.B.; Vijayaraghavan, P.K.
    Stability of ß-carotene in isolated systems.
    Journal of Food Technology 14 (-) 571-578; 1979
    4
76 Arya, S.S.; Natesan, V.; Pavihar, D.B.; Vijayaraghavan, P.K.
    Stability of carotenoids in dehydrated carrots.
    Journal of Food Technology 14 (-) 579-586; 1979
    4,5
77 Ashpole, D.K.
    The moisture relations of textile fibres at high humidities.
    Proc. of the Royal Society London A 212, 112-123; 1952
    1,2,3
78 Asselmeyer, F.; Zott, H.
    Sorption von Wasserdampf an NaCl-Oberflächen.
    Zeitschrift für angewandte Physik 19 (3) 168-175; 1965
    2,3,5
79 Astakhov, V.A.; Lepilin, V.N.; Romankov, P.G.; Lukin, V.D.
    Equation for an isotherm of the adsorption of vapors by
    activated carbons with microporous structures.(Orig.russ.)
    Zhurnal Prikldnoi Khimii 41 (6) 1240-1245; 1968
    2
80 Attrey, D.P.; Sharma, T.R.
    Sorption isotherms and monolayer moisture content of raw
    freeze dried mutton.
    Journal of Food Science and Technology, India 16 (4) 155-158;
    1979
    5
81 Audu, T.O.K.; Loncin, M.; Weisser, H.
    Sorption isotherms of sugars.
    Lebensmittel-Wissenschaft und -Technologie 11 (1) 31-34; 1978
    3,5
82 Auker, C.A.; Geddes, W.F.; Bailey, C.H.
    A study of the net weight changes and moisture content of
    wheat flour at various relative humidities.
    Cereal Chemistry 19 (-) 128; 1942
    2,4

83 Ayernor, G.S.
Particulate properties and rheology of pregelled yam
(dioscorea rotundata) products.
Journal of Food Science 41 (-) 180-182; 1976
4,5
84 Ayernor, G.S.; Steinberg, M.P.
Hydration and rheology of soy-fortified pregelled corn flours.
Journal of Food Science 42 (1) 65-69; 1977
4,5
85 Ayerst, G.
Wateractivity, its measurement and significance in biology.
International Biodeterioration Bulletin 1 (2) 13-26; 1965
1,3,4
86 Ayerst, G.
Determination of the wateractivity of some hygroscopic food
materials by a dew point method.
Journal of Science of Food and Agriculture 16 (2) 71-78; 1965
3,5
87 Ayerst, G.
Wateractivity - its measurement and significance in biology -
a review article.
Food Technology in Australia 20 (2) 80-86; 1968
1,3,4
88 Ayerst, G.
The effects of moisture and temperature on growth and spore
germination in some fungi.
Journal of Stored Products Research 5 (2) 127-141; 1969
4,5
89 Babbitt, J.D.
On the adsorption of water vapour by cellulose.
Canadian Journal of Research 20 A (-) 143-172; 1942
2,5
90 Babbitt, J.D.
Hysteresis in the adsorption of water vapor by wheat.
Nature 156 (-) 265-266; 1945
2,5
91 Babbitt, J.D.
Observations on the adsorption of water vapor by wheat.
Canadian Journal of Research 27 F (-) 55-72; 1949
2,3,5
92 Bach, R.O; Boardman, W.W. jr.
Vapor pressure of aqueous lithium iodide solutions.
ASHRAE Journal 9 (11) 33-36; 1967
3
93 Bachelet, M.; Huguet, J.
Utilisation des ultrasons pour la mesure des pressions de vapeur
de produits solides hydratés.
Comptes Rendus de l'Academie des Sciences 262 (-) 1308-1310;
1966
3
94 Bacon, C.; Sweeney, J.; Robbins, J.; Burdick, D.
Production of pencillic acid and ochratoxin A on poultry feed
by Aspergillus ochraceus: temperature and moisture requirements.
Applied Microbiology 26 (2) 155-160; 1973
4
95 Baddiel, C.; Breuer, M.; Stephens, R.
The mechanism of water sorption in poly-L-alanine.
Journal of Colloid and Interface Science 40 (3) 429-436; 1975
2

96 Bahner, F.; Motika, W.
    Feuchtemeßgerät für hohe Temperaturen und hohe Feuchtegrade.
    Chemie-Ingenieur-Technik 45 (10a) 732-736; 1973
    3
97 Bahrs, L.W.; Herrero, F.M.
    Neue Methode zur Messung der Wasseraktivität.
    Die Ernährungswirtschaft 26 (3) 39-40; 1979
    3,4
98 Baiker, A.; Richarz, W.
    Vergleich von Meßmethoden zur Bestimmung der
    Porenradienverteilung und spezifischen Oberfläche poröser
    Katalysatoren.
    Chemie-Ingenieur-Technik 49 (5) 399-403; 1977
    3
99 Bailey, C.H.
    The hygroscopic moisture of flour exposed to atmospheres of
    different relative humidity.
    Journal of Industrial Engineering Chemistry 12 (11) 1102-1104;
    1920
    5
100 Bailey, C.H.
    Respiration of shelled corn.
    Techn. Bull. 3. The University of Minnesota, Agricultural
    Experiment Station; 1921
    5
101 Baird-Parker, A.; Boothroyd, M.; Jones, E.
    The effect of wateractivity on the heat resistant strains of
    salmonellae.
    Journal of Applied Bacteriology 33 (3) 512-522; 1970
    4
102 Baird-Parker, A.C.; Freame, B.
    Combined effect of water activity, pH and temperature on the
    growth of Clostridium botulinum from spore and vegetative cell
    inocula.
    Journal of Applied Bacteriology 30 (-) 420-429; 1967
    4
103 Baisya, R.K.; Chattoraj, D.K. Bose, A.N.
    Water binding characteristics of normal curd (dahi) and
    rehydrated curd powder gels.
    Indian Journal of Technology 13 (12) 578-581; 1975
    5
104 Baisya, R.K.; Chattoraj, D.K.; Bose, A.N.
    Studies on the physico-chemical characteristics of fresh curd
    (dahi) and rehydrated freeze dried curd powder gels.
    Journal of Food Science and Technology, India 15 (2) 71-75;
    1978
    2,5
105 Bakharev, I.Y.
    Cited in: Gerzhoi, A.P.; Samochetov, V.F.
    Grain Drying and Grain Dryers.
    Published for the National Science Foundation and Department
    of Agriculture by the Israel Program for Scientific
    Translations; 1960
    5
106 Ballschmieter, H.M.B.
    Vergleichende Wasserbestimmung von Lebensmitteln nach
    verschiedenen Methoden.
    Deutsche Lebensmittel-Rundschau 63 (7) 203-207; 1967
    3

107 Ballschmieter, H.M.B.
Die Sorptionsisothermen von Mehlen, Mehlprodukten und anderen
Lebensmitteln bei 30 °C.
Getreide und Mehl 17 (10) 118-120; 1967
5

108 Baloch, A.K.; Buckle, K.A.; Edwards, R.A.
Stability of ß-carotene in model systems containing sulphite.
Journal of Food Technology 12 (3) 309-316; 1977
4,5

109 Bandyopadhyay, S.; Weisser, H.; Loncin, M.
Water adsorption isotherms of foods at high temperatures.
Lebensmittel-Wissenschaft und -Technologie 13 (4) 182-185;
1980
3,5

110 Banks, H.J.
Effects of controlled atmosphere storage on grain quality:
A review.
Food Technology in Australia 33 (7) 335-340; 1981
4

111 Barbetti, P.
Effeto del grado di umidita sulla reazione di Maillard. Studio
di alcuni sistemi-modello allo stato condensato.
Industria Alimentaria 15 (128) 94-98; 1976
4

112 Baresel, D.; Gellert, W.
Eine einfache Apparatur zur Bestimmung der
Porenradien-Verteilung in porösen Körpern.
Chemie-Ingenieur-Technik 43 (3) 128-131; 1971
3

113 Baret, J.F.
Theoretical model for an interface allowing a kinetic study of
adsorption.
Journal of Colloid and Interface Science 30 (1) 1-12; 1969
2

114 Barkas, W.W.
The swelling of wood under stress.
Her Majesty's Stationery Office (H.M.S.O.), London; 1949
2

115 Barr, M.; Tice, L.
A study of inhibitory concentration of various sugar and
polyols on the growth of microorganisms.
Journal of the American Pharmaceutical Association
46 (4) 219-221; 1957
4

116 Barraud, C.; Grimault, M.L.
L'activité de l'eau aw des produits a base de viande, salaison
et charcuterie.
Industries Alimentaires et Agricoles 97 (1, 2) 17-29; 1980
2,3,4

117 Barraud, C.; Grimault, M.L.
Water activity (aw) in cooked and uncooked meat products and
in pickled products.
EG-Doc.III/420/1980;Orig.Fr.,Cost 90/6-80
1,2,3,4

118 Barrer, R.M.; McKenzie, N.; Reay, J.S.S.
Capillary condensation in single pores.
Journal of Colloid Science 11 (-) 479-495; 1956
2

119 Barrie, J.A.
Diffusion in polymers.
In: Crank, J.; Park, G.S. (Edts.).
Academic Press, London, New York; 1968
5

120 Barrie, J.A.; Machin, D.
Diffusion and association of water in some
polyalkylmethacrylades.
Part 1: Equilibrium sorption and steady state permeation.
Transactions of the Faraday Society 67 (-) 244-256; 1971
2

121 Barron, L.F.
The expansion of wafer and its relation to the cracking of
chocolate and "bakers" chocolate coatings.
Journal of Food Technology 12 (1) 73-84; 1977
4,5

122 Bartholomai, G.B.; Brennan, J.G.; Jowitt, R.
Mechanisms of volatile retention in freeze drying. The
contributon of adsorption.
Proc. of the IV International Congress of Food Science
and Technology. Vol. IV, pp 169-174; 1974
2

123 Bartholomai, G.B.; Brennan, J.G.; Jowitt, R.
Mechanisms of volatile retention in freeze dried food liquids.
Lebensmittel -Wissenschaft und -Technologie 8 (1) 25-28; 1975
2,5

124 Bartolini, R.; Lerici, C.
I.M.F.: Prodotti alimentari ad umidita intermedia.
Rivista di Scienza e Tecnologia degli Alimenti e di Nutrizione
Umana 6 (4) 243-248; 1976
4

125 Barton-Wright, E.C.
Studies on the storage of wheaten flour.
III. Changes in the flora and the fats and the influence of
these changes on gluten character.
Cereal Chemistry 15 (-) 521-541; 1938
4

126 Bas, E.B.
Desorptionsspektrometrie bei tiefen Temperaturen.
Vakuum-Technik 14 (3) 65-69; 1965
2

127 Basler, W.; Lechert, H.
Kernresonanzuntersuchungen an Wasser in Stärkegelen bei 295 K.
Die Stärke 25 (9) 289-292; 1973
2

128 Basler, W.; Lechert, H.
Diffusion von Wasser in Stärkegelen.
Die Stärke 26 (2) 39-42; 1974
2

129 Bates, D.
Practical aspects of relative humidity measurement.
In: Relative humidity in the food industry.
British Food Manufacturing Industries Research Association
(B.F.M.I.R.A.). Symposium Proc. No. 4, 9-15, Leatherhead,
London; 1969
3

130 Baumann, F.
Apparatur nach Baumann zur Bestimmung der Flüssigkeitsaufnahme
von pulverigen Substanzen.
Glas- und Instrumententechnik Fachzeitschrift für das
Laboratorium 11 (6) 540-542; 1967
2,3

131 Baumgardt, J.P.
Qualitätsbeeinflussung von Dauerbackwaren.
Brot und Gebäck 14 (10) 189-194; 1960
4,5

132 Baunack, F.
     Die Trocknung von Stärke.
     Die Stärke 9 (8) 143-146; 1957
     2,5
133 Baunack, F.
     Über den Wärmeverbrauch bei der Trocknung.
     Zucker 13 (16) 400-405; 1960
     2,5
134 Baunack, F.; Kessenich, A.
     Über die Schrumpfung und Quellung von ausgelaugten
     Zuckerrübenschnitzel.
     Zucker 14 (23) 599-606; 1961
     2,5
135 Bawa, A.S.; Manjrekar, S.P.
     Effect of water activity on quality of accelerated
     freeze-dried mutton.
     Indian Food Packer 31 (6) 53-56; 1977
     4
136 Baxter, G.P.; Cooper, W.C. jr.
     The aqueous pressure of hydrated crystals. II Oxalic acid,
     sodium sulfate, sodium acetate, sodium carbonate, disodium
     phosphate, barium chloride.
     Journal of the American Chemical Society 46 (-) 923; 1924
     3
137 Baxter, G.P.; Lausing, J.E.
     The aqueous pressure of some hydrated crystals. Oxalic acids,
     strontium chloride and sodium sulfate.
     Journal of the American Chemical Society 42 (-) 419; 1920
     3
138 Beary, E.G.
     Moisture equilibrium in relation to the chemical stability of
     dehydrated foods. A bibliography.
     Technical Library U.S. Army Natick Laboratories Natick, Mass.
     Bibliographic Series 67-1, No. 20; 1967
     4
139 Beasley, E.O.
     Moisture equilibrium of Virginia Bunch peanuts.
     M.S. Thesis, North Carolina State University, Raleigh, N.C.;
     1962
     3,5
140 Beattie, M.V.F.
     Observations of the thermal death points of the blowfly at
     different relative humidities.
     Bulletin of Entomological Research 18 (-) 397-403; 1928
     4
141 Becht, W.A.
     Water binding by starch-lecithin system for intermediate
     moisture foods.
     M.S. Thesis, University of Illinois, Urbana; 1977
     4
142 Bechtel, P.J.; Palnitkar, M.P.; Heldmann, D.R.; Pearson, A.M.
     Bound water determination using vacuum differential scanning
     calorimetry.
     Journal of Food Science 36 (3) 84-86; 1971
     2
143 Beck, G.; Mantel, T.
     Beurteilungskriterien der Schnittfestigkeit von Rohwürsten.
     Fleischwirtschaft 57 (2) 243-245; 1977
     4,5

144 Beckel, A.C.; Cartter, J.L.
The effect of variety and environment on the equilibrium
moisture content of soybean seed.
Cereal Chemistry 20 (-) 362-368; 1943
2,5

145 Becker, H.A.
An interpretation of the moisture desorption isotherm of
wheat.
Canadian Journal of Chemistry 36 (-) 1416-1423; 1958
5

146 Becker, H.A.
On the adsorption of liquid water by the wheat kernel.
Cereal Chemistry 37 (5) 309-323; 1960
2

147 Becker, H.A.; Sallans, H.R.
A study of the desorption isotherms of wheat at 25°C and 50°C.
Cereal Chemistry 33 (2) 79-91; 1956
2,5

148 Becker, H.A.; Sallans, H.R.
Theoretical study of the mechanism of moisture diffusion in
wheat.
Cereal Chemistry 34 (6) 395-409; 1957
2

149 Beever, D.K.; Valentine, L.
Studies on the sorption of moisture by polymers. II. The
cellulose-cellulose acetate system.
Journal of Applied Chemistry 8 (-) 103-107; 1958
5

150 Belchev, N.
Water activity in foodstuffs.
Konservna Promishlenost - (10) 19-22; 1981
1

151 Beleau, M.H.; Heidelbaugh, N.D.; Dyke, D. van
Open-ocean farming of kelp.
Food Technology 27 (12) 27-30, 45; 1975
5

152 Bem, Z.; Leistner, L.
Bestimmung der Wasseraktivität von Fleisch und Fleischwaren
mit der Methode von Landrock und Proctor.
Fleischwirtschaft 50 (10) 1412-1414; 1970
3,5

153 Bem, Z.; Leistner, L.
Die Wasseraktivitätstoleranz der bei Pökelfleischwaren
vorkommenden Hefen.
Fleischwirtschaft 50 (4) 492-493; 1970
4

154 Bemiller, J.N.; Pratt, G.W.
Sorption of water, sodium sulfate and water soluble alcohols
by starch granules in aqueous suspension.
Cereal Chemistry 58 (6) 517-520; 1981
2

155 Bender, E.; Block, U.
Bedeutung der Mischphasen-Thermodynamik für die thermische
Trenntechnik aus der Sicht der chemischen Industrie.
Chemie-Ingenieur-Technik 49 (6) 479-487; 1977
2

156 Benmergui, E.A.; Ferro-Fontan, C.; Chirife, J.
The prediction of water activity in aqueous solutions in
connection with intermediate moisture foods.
I. aw prediction in single aqueous electrolyte solutions.
Journal of Food Technology 14 (6) 625-637; 1979
2,4

157 Benson, S.W.; Ellis, D.A.
     Suface areas of proteins.
     I. Suface areas and heats of absorption.
     Journal of the American Chemical Society 70 (-) 3563-3569;
     1948
     2,3
158 Benson, S.W.; Ellis, D.A.
     Surface areas of proteins.
     II. Adsorption of non polar gases.
     Journal of the American Chemical Society 72 (-) 2095-2102;
     1950
     2,3
159 Benson, S.W.; Ellis, D.A.; Zwanzig, R.W.
     Surface areas of proteins.
     III. Adsorption of water.
     Journal of the American Chemical Society 72 (-) 2102-2105;
     1950
     2,3
160 Benson, S.W.; Richardson, R.L.
     A study of hysteresis in the sorption of polar gases by native
     and denatured proteins.
     Journal of the American Chemical Society 77 (-) 2585-2590;
     1955
     2,5
161 Benson, S.W.; Seehof, J.M.
     The surface areas of proteins.
     IV. Sorption of polar gases.
     Journal of the American Chemical Society 73 (-) 5053-5058;
     1951
     2
162 Benson, S.W.; Srinivasan, R.
     The effect of temperature on the sorption of polar gases by
     proteins.
     Journal of the American Chemical Society 77 (-) 6371-6372;
     1955
     2
163 Berendsen, H.J.C.
     Nuclear magnetic resonance study of collagen hydration.
     Journal of Chemical Physics 36 (-) 3297; 1962
     2
164 Berendsen, H.J.C.
     Specific interactions of water with biopolymers.
     In: Franks, F.(Ed.). Water A Comprehensive Treatise.
     Vol. 5, pp 293-349. Water in Disperse Systems.
     Plenum Press, New York; 1975
     1
165 Berger, M.A.; Richard, N.; Cheftel, J.C.
     Influence de l'activité de l'eau sur la thermorésistance d'une
     levure osmophile: Saccharomyces rouxii, dans un aliment à
     humidité intermédiaire.
     Lebensmittel -Wissenschaft und -Technologie 15 (2) 83-88; 1982
     4
166 Berlin, E.
     Hydration of milk proteins.
     In: Rockland, L.B.; Stewart, G.F. (Edts.).
     Water acitivity: Influences on Foodquality. pp 467-488.
     Academic Press, New York; 1981
     2,5
167 Berlin, E.; Anderson, B.A.
     Reversibility of water vapor sorption by cottage cheese whey
     solids.
     Journal of Dairy Science 58 (1) 25-29; 1975
     5

168 Berlin, E.; Anderson, B.A.; Pallansch, M.J.
    Effect of water vapor sorption on porosity of dehydrated dairy
    products.
    Journal of Dairy Science 51 (5) 668-672; 1968
    4
169 Berlin, E.; Anderson, B.A.; Pallansch, M.J.
    Water vapor sorption properties of various dried milks and
    wheys.
    Journal of Dairy Science 51 (9) 1339-1344; 1968
    3,5
170 Berlin, E.; Anderson, B.A.; Pallansch, M.J.
    Comparison of water vapor sorption by milk powder components.
    Journal of Dairy Science 51 (12) 1912-1915; 1968
    5
171 Berlin, E.; Anderson, B.A.; Pallansch, M.J.
    Effect of temperature on water vapor sorption by dried milk
    products.
    Journal of Dairy Science 52 (6) 884; 1969
    2,5
172 Berlin, E.; Anderson, B.A.; Pallansch, M.J.
    Morphological alterations in dried casein particles effected
    by sequential water vapor sorption-desorption cycles.
    Journal of Dairy Science 52 (6) 884-885; 1969
    4
173 Berlin, E.; Anderson, B.A.; Pallansch, M.J.
    Sorption of water vapor and of nitrogen by genetic variants of
    alpha s1-casein.
    Journal of Physical Chemistry 73 (2) 303-307; 1969
    2,5
174 Berlin, E.; Anderson, B.A.; Pallansch, M.J.
    Effect of temperature on water vapor sorption by dried milk
    powders.
    Journal of Dairy Science 53 (2) 146-149; 1970
    2,5
175 Berlin, E.; Anderson, B.A.; Pallansch, M.J.
    Contraction of dried casein particle surfaces effected by
    sequential water vapor sorption and desorption cycles.
    Journal of Colloid and Interface Science 33 (2) 312; 1970
    5
176 Berlin, E.; Anderson, B.A.; Pallansch, M.J.
    Effect of hydration and crystal form on the surface area of
    lactose.
    Journal of Dairy Science 55 (10) 1396-1399; 1972
    5
177 Berlin, E.; Anderson, B.A.; Pallansch, M.J.
    Water sorption by dried dairy products stabilized with
    carboxymethyl cellulose.
    Journal of Dairy Science 56 (6) 685-689; 1973
    Abstracts of Papers. American Chemical Society 162 AGFD 47;
    1971
    2,5
178 Berlin, E.; Anderson, B.A.; Pallansch, M.J.
    Influence of dehydration method on the adsorption of benzene
    vapor by dried casein.
    Journal of Colloid and Interface Science 43 (3) 571-576; 1973
    3
179 Berlin, E.; Howard, N.M.; Pallansch, M.J.
    Specific surface areas of milk powders produced by different
    drying methods.
    Journal of Dairy Science 47 (2) 132-138; 1964
    3

180 Berlin, E.; Kliman, P.G.; Anderson, B.A.; Pallansch, M.J.
     Calorimetric measurement of the heat of desorption of water
     vapor from amorphous and crystalline lactose.
     Thermochimica Acta 2 (-) 143-152; 1971
     2
181 Berlin, E.; Kliman, P.G.; Anderson, B.A.; Pallansch, M.J.
     Water binding in whey protein concentrates.
     Journal of Dairy Science 56 (8) 984-987; 1973
     2,5
182 Berlin, E.; Kliman, P.G.; Pallansch, M.J.
     Surface areas and densities of freeze-dried foods.
     Journal of Agricultural and Food Chemistry 14 (1) 15-17; 1966
     3
183 Berlin, E.; Kliman, P.G.; Pallansch, M.J.
     Water binding by milk powder components - calorimetric study.
     Journal of Dairy Science 52 (6) 884; 1969
     2
184 Berlin, E.; Kliman, P.G.; Pallansch, M.J.
     Changes in state of water in proteinaceous systems.
     Journal of Colloid and Interface Science 34 (4) 488-494; 1970
     2
185 Berlin, E.; Kliman, P.G.; Pallansch, M.J.
     Effect of sorbed water on the heat capacity of crystalline
     proteins.
     Thermochimica Acta 4 (-) 11-16; 1972
     4
186 Bernal, J.D.; Fowler, R.H.
     A theory of water and ionic solution, with particular
     reference to hydrogen and hydroxyl ions.
     Journal of Chemical Physics 1 (8) 515-548; 1933
     2
187 Bernett, M.K.; Zisman, W.A.
     Effect of adsorbed water on wetting properties of borosilicate
     glass, quartz, and sapphire.
     Journal of Colloid and Interface Science 29 (3) 413-423; 1969
     4
188 Berry, M.R. jr.; Dickerson, R.W. jr.
     Moisture adsorption isotherms for selected feeds and
     ingredients.
     Transactions of the ASAE 16 (1) 137-139; 1973
     2,5
189 Best, R.; Spingler, E.
     Messung von Adsorptions- und Desorptionsisothermen mit einer
     vollautomatischen Apparatur.
     Chemie-Ingenieur-Technik 44 (21) 1222-1226; 1972
     3
190 Best, S.M.; Hullet, E.W.
     The equilibrium moisture relations of New Zealand wheat.
     New Zealand Journal of Science 11 (3) 97-104; 1968
     4,5
191 Bettelheim, F.A.
     The calculation of thermodynamic quantities from hysteresis
     data.
     Journal of Colloid and Interface Science 23 (-) 297-301; 1967
     2
192 Bettelheim, F.A.
     Comments on Professor La Mer's Letter.
     Journal of Colloid and Interface Science 23 (-) 301; 1967
     2

193 Bettelheim, F.A.; Block, A.
Water vapor sorption of bovine and porcine submaxillary
mucins.
Biochimica et Biophysica Acta 165 (-) 405-409; 1968
5

194 Bettelheim, F.A.; Block, A.; Kaufmann, L.J.
Heats of water vapor sorption in swelling biopolymers.
Biopolymers 9 (-) 1531-1538; 1970
2

195 Bettelheim, F.A.; Ehrlich, S.H.
Water vapor sorption of mucopolysaccharides.
Journal of Physical Chemistry 67 (-) 1948-1953; 1963
2,5

196 Bettelheim, F.A.; Sterling, C.; Volman, D.H.
Pectic substances-water. I. Structural changes in the
polygalacturonic chains during water adsorption.
Journal of Polymer Science 22 (-) 303-314; 1956
3,5

197 Bettelheim, F.A.; Volman, D.H.
Pectic substances-water. II. Thermodynamics of water vapor
sorption.
Journal of Polymer Science 24 (-) 445-454; 1957
2

198 Bettelheim, F.A.; Volman, D.H.
The sorption of water vapor by calcium pectate.
Journal of Polymer Science 24 (107) 485-488; 1957
5

199 Beuchat, L.R.
Combined effects of water activity, solute and temperature on
the growth of Vibrio parahaemolyticus.
Applied Microbiology 27 (6) 1075-1080; 1974
4

200 Beuchat, L.R.
Relationship of water activity to moisture content in tree
nuts.
Journal of Food Science 43 (3) 754-755, 758; 1978
2,5

201 Beuchat, L.R.
Food and beverage mycology.
AVI Publ. Comp. Inc. Westport, Conn.; 1978
4

202 Beuchat, L.R.
Influence of water activity on growth, metabolic activities
and survival of yeasts and molds.
Journal of Food Protection 46 (2) 135-141; 1983
4

203 Beuchat, L.R.; Heaton, E.K.
Factors influencing fungal quality of pecans stored at
refrigeration temperatures.
Journal of Food Science 45 (-) 251-254; 1980
4

204 Beuchat, L.R.; Lechowich, R.V.
Aflatoxins: Production on beans as affected by temperature
and moisture content.
Journal of Milk and Food Technology 33 (-) 373-376; 1970
4

205 Beuschel, H.
Untersuchungen über die Ursachen und
Beeinflussungsmöglichkeiten der Schwindspannungen bei der
Trocknung pastenartiger Stoffe, insbesondere von Teigwaren.
Dissertation, TH München; 1955
3,4,5

206 Beutler, H.O.
Über den Einfluß der relativen Luftfeuchtigkeit auf
enzymatische Reaktionen in Getreidemahlprodukten unter
besonderer Berücksichtigung der Bildung von Säuren.
Dissertation, Universität Münster/Westfalen; 1963
4
207 Bevan, S.C.; Gregg, S.J.; Parkyns, N.D. (Edts.).
Progress in vacuum microbalance techniques. Vol. 2,
Heyden and Son Ltd., London; 1973
3
208 Beyer, H.; Schay, G.
Bestimmung von Adsorptionsisothermen und Oberflächengrößen
nach einer modifizierten Sorptometermethode.
Gaschromatographie (-) D155-D162; 1965
3
209 Bienenstock, B.; Powers, H.E.C.
Introducing the "Equilibrium Relative Humidity" of a sugar.
The International Sugar Journal 53 (-) 254-255; 1951
2
210 Bimbenet, J.J.; Brusset, H.; Loncin, M.
Effets de la présence de corps solubles sur la deshydratation
des produits biologiques.
Industries Alimentaires et Agricoles 87 (4) 385-391; 1970
2,5
211 Bimbenet, J.J.; Guilbot, A.
Modifications biochimiques et physicochimiques au cours du
sèchage.
Chimie et Industrie, Genie Chimique 96 (4) 1-12; 1966
2
212 Bimbenet, J.J.; Le Maguer, M.; Loncin, M.
Les phénomènes de sorption. Applications à la lyophilisation.
Annexe 1969-9 au Bulletin de l'Institut International du Froid.
pp 21-39. Paris; 1969
1,2,3,5
213 Bindon, H.H.
A critical review of tables and charts used in psychrometry.
In: Wexler, A.; Ruskin, R.E. (Edts.).
Humidity and Moisture; Measurement and Control in Science and
Industry.
Vol. 1, pp 3-15. Reinhold Publ. Corporation, New York; 1965
3
214 Biran, D.; Giacin, J.R.; Hayakawa, K.; Gilbert, S.G.
Vinylchloride sorption by dry casein particles: mechanistic
considerations.
Journal of Food Science 44 (1) 59-61; 1979
2
215 Biran, D.; Gilbert, S.G.; Giacin, J.R.
Sorption of vinylchloride by selected food constituents.
Journal of Food Science 44 (1) 56-58; 1979
1
216 Bizot, H.
Using the 'GAB' model to construct sorption isotherms.
In: Jowitt, R.; Escher, F.; Hallström, B.; Meffert, H.F.T.;
Spieß, W.E.L.; Vos, G. (Edts.). Physical Properties of Foods.
pp 43-53.
Applied Science Publ., London, New York; 1983
2

217 Bizot, H.; Buleon, A.; Delage, M.M.; Multon, J.L.
     Influence de l'hystérésis de sorption de la vapeur d'eau, sur
     les diagrammes de diffraction "X" et la masse volumique de
     l'amidon de pomme de terre.
     Science des Aliments 1 (3) 401-413; 1981
     2,4
218 Bizot, H.; Multon, J.L.
     Méthode de référence pour la mesure de l'activité de l'eau
     dans les produits alimentaires.
     Annales de Technologie Agricole 27 (2) 441-454; 1978
     1,3
219 Bizot, H.; Tome, D.; Guilbot, A.; Drapron, R.; Multon, J.L.
     Les aliments à humidité intermédiaire.
     Série Synthèse bibliographique, No. 16, CDIUPA et APRIA,
     Paris; 1978
     4
220 Bizot, J.
     Méthode automatique de dosage coulométrique de petites
     quantités d'eau.
     Bulletin de la Société Chimique de France - (1) 151-157; 1967
     3
221 Blain, J.
     Moisture levels and enzyme activity.
     In: Hawthorn, J.; Leitch, M.J.(Edts.). Recent Advances in Food
     Science. Vol. 2, Processing. pp 41-45. Butterworth, London; 1962
     4
222 Bloch, F.; Brekke, J.E.
     Processing of pistachio nuts.
     Economic Botany 14 (2) 129-144; 1960
     5
223 Block, A.; Bettelheim, F.
     Water vapor sorption of hyaluronic acid.
     Biochemica et Biophysica Acta 201 (1) 69-75; 1970
     5
224 Block, A.; Hewitt, E.J.
     Study of the application of relative humidity and moisture
     vapor pressure measurements for the determination of the
     moisture content of dehydrated foods.
     Evans Research and Development Corp. Progress Reports
     15 February - 15 November 1963, Nos. 1/2, No 3.
     1
225 Block, S.S.
     Humidity requirements for mold growth.
     Applied Microbiology 1 (6) 287-293; 1953
     4,5
226 Block, S.S.; Rodriguez-Torrent, R.; Cole, M.B.; Prince, A.E.
     Humidity and temperature requirements of selected fungi.
     Developments in Industrial Microbiology 3 (-) 204-216; 1962
     4
227 Bluestein, P.M.
     The kinetics of sorption of water vapor in freeze-dried food.
     Ph.D. Thesis, Mass. Inst. Technol. Cambridge, MA, USA; 1971
     2,3,5
228 Bluestein, P.M.; Labuza, T.P.
     Kinetics of water vapor sorption in a model freeze-dried food.
     American Institute of Chemical Engineers Journal 18 (4) 706-712;
     1972
     2,3

229 Boehm, H.P.; Sappok, R.
Microgravimetric studies of water vapor adsorption on solid surface.
In: Gast, T.; Robens, E. (Edts.).
Progress in Vacuum Microbalance Techniques. Vol. 1, pp 247-264
Heyden and Son, Ltd., London; 1972
3

230 Böhme, G.; Robens, E.; Straubel, H.; Walter, G.
Determination of relative weight changes of electrostatically suspended particles in the sub-microgram range.
In: Bevan, S.C.; Gregg, S.J.; Parkyns, N.D. (Edts.).
Progress in Vacuum Microbalance Techniques. Vol. 2, pp 169-174
Heyden and Son, Ltd., London; 1973
3

231 Boggs, M.M.; Fevold, H.L.
Dehydrated egg powders. Factors in palatability of stored powders.
Industrial Engineering Chemistry 38 (-) 1075-1079; 1946
4

232 Bogsanyi, D.J.; Weeden, D.G.
A simple apparatus for water vapor pressure-moisture studies by a dew point method.
Chemistry and Industry 23 (-) 741-742; 1968
3

233 Bolin, H.R.
Relation of moisture to water activity in prunes and raisins.
Journal of Food Science 45 (5) 1190-1192; 1980
3,5

234 Boller, R.A.; Schroeder, H.W.
Influence of relative humidity on production of aflatoxin in rice by Aspergillus parasiticus.
Phytopathology 64 (-) 17-21; 1974
4

235 Bolliger, W.; Gál, S.; Signer, R.
Eine Apparatur zur automatischen Bestimmung von Wasserdampf-Sorptions-Isothermen und -Isobaren.
Helvetica Chimica Acta 55 (7) 2659-2663; 1972
3

236 Bone, D.P.
Water activity - its chemistry and applications.
Food Product Development 3 (5) 81, 84-85, 88, 90, 92, 94; 1969
1

237 Bone, D.P.
Water activity in intermediate moisture foods.
Food Technology 27 (4) 71-72, 74, 76; 1973
1,4

238 Bone, D.P.; Ross, K.D.; Shannon, E.L.; Flanyak, J.R.
Some factors in formulating and processing intermediate moisture foods.
Proc. of the American Association of Cereal Chemists, Montreal; Oct. 1974
4

239 Bone, D.P.; Shannon, E.L.; Ross, K.D.
The lowering of water activity by order of mixing in concentrated solutions.
In: Duckworth, R.B. (Ed.). Water Relations of Foods.
pp 613-626. Academic Press, London; 1975
2,4

240 Bone, S.; Gascoyne, P.R.C.; Pethig, R.
Dielectric properties of hydrated proteins at 9.9 GHZ.
Journal of the Chemical Society: Faraday Trans I,
73 (-) 1607-1611; 1977
3,5

241 Bonilla-Vidales, C.M.; Plett, E.A.; Loncin, M.
    Untersuchungen zur Aromaerhaltung während der Lufttrocknung
    von Zuckerlösungen.
    Lebensmittel-Wissenschaft und -Technologie 14 (3) 153-159; 1981
    2
242 Bonner, J.T.
    A study of the temperature and humidity requirements of
    Aspergillus niger.
    Mycologia 40 (-) 728-738; 1948
    4
243 Boquet, R.; Chirife, J.; Iglesias, H.A.
    A criterium for the evaluation of significant differences
    between water sorption isotherms of a food sample subjected to
    different treatments.
    Lebensmittel-Wissenschaft und -Technologie 10 (5) 246-248;
    1977
    2
244 Boquet, R.; Chirife, J.; Iglesias, H.A.
    Equations for fitting water sorption isotherms of foods.
    II. Evaluation of various two-parameter models.
    Journal of Food Technology 13 (4) 319-327; 1978
    2
245 Boquet, R.; Chirife, J.; Iglesias, H.A.
    Equations for fitting water sorption isotherms of foods.
    III. Evaluation of various three-parameter models.
    Journal of Food Technolgy 14 (5) 527-534; 1979
    2
246 Boquet, R.; Chirife, J.; Iglesias, H.A.
    On the equivalence of isotherm equations.
    Journal of Food Technology 15 (3) 345-349; 1980
    2
247 Borukh, I.F.; Panchenko, G.N.
    Hygroscopic properties and equilibrium moisture content in
    honey. (Orig.russ.)
    Khlebopekarnaya i Konditerskaya Promyshlennost' 11, 22-23;
    1973
    5
248 Bosin, W.A.; Easthouse, H.D.
    Rapid method for obtaining humidity equilibrium data.
    Food Technology 24 (10) 1155-1157; 1970
    3,5
249 Boskovis, M.A.
    Water sorption properties and storage stability of foam-mat
    dried tomato powder.
    First International Congress on Engineering and Food, August
    9-13, Boston; 1976
    4,5
250 Bourland, C.T.; Rapp, R.M.; Smith, M.C.
    Space shuttle food system.
    Journal of Food Technology 31 (9) 40-41, 44-45; 1977
    4
251 Bourrie, G.; Pedro, G.
    The pF concept, its physicochemical basis and pedogenetic
    significance.
    I. Physicochemical aspect. Relationships between pF and
    activity of water.
    Bulletin de l'Association Francaise pour l'Etude du Sol
    4 (-) 313-322; 1979
    4

252 Bousquet-Ricard, M.
    Aliments a humidité intermédiaire.
    Thèse, Académie de Montpellier Université des Sciences
    et Techniques du Languedoc; 1979
    2,4,5

253 Bousquet-Ricard, M.; Quayle, G.; Pham, T.; Cheftel, J.C.
    Etude comparative critique de trois méthodes de mesure de
    l'activité de l'eau des aliments a humidité intermédiaire.
    Lebensmittel-Wissenschaft und -Technologie 13 (4) 169-176; 1980
    3

254 Box, J.E. jr.
    Design and calibration of a thermocouple psychrometer which
    uses the peltier effect.
    In: Wexler, A.; Ruskin, R.E. (Edts.).
    Humidity and Moisture. Measurement and Control in Science and
    Industry. Vol. 1, pp 110-121. Reinhold Publ. Corporation,
    New York; 1965
    3

255 Boyer, J.S.
    Leaf water potentials measured with a pressure chamber.
    Plant Physiology 42 (-) 133-137; 1967
    3

256 Boylan, S.L.; Acott, K.M.; Labuza, T.P.
    Staphylococcus aureus F 265 challenge studies in an
    intermediate moisture food.
    NASA Report, Phase III; 1975
    4

257 Boylan, S.L.; Acott, K.M.; Labuza, T.P.
    Staphylococcus aureus challenge study in an intermediate
    moisture food.
    Journal of Food Science 41 (4) 918-921; 1976
    4

258 Bradley, R.S.
    Polymer adsorbed films. Part I. The adsorption of argon on salt
    crystals at low temperatures, and the determination of surface
    fields.
    Journal of the Chemical Society 58 (-) 1467-1474; 1936
    3

259 Bradley, R.S.
    Polymolecular adsorbed films. Part II. The general theory of
    the condensation of vapors on finely divided solids.
    Journal of the Chemical Society 58 (-) 1799-1804; 1936
    2

260 Brandenburg, N.R.; Harmond, J.E.
    Equilibrium moisture content of fiber flax.
    U.S. Dept. Agr. Bull. 1200 (1959)
    3,5

261 Brandt, W.W.; Budrys, R.S.
    Sorption rates indicative of structural changes in solid
    polypeptides.
    Journal of Physical Chemistry 69 (2) 600-604; 1965
    2,4,5

262 Brannen, J.P.
    Role of water activity in the dry heat sterilization of micro-
    organisms.
    Journal of Theoretical Biology 32 (-) 331; 1971
    4

263 Brannen, J.P.; Garst, D.
    Dry heat inactivation of Bacillus subtilis var. Niger spores a
    a function of relative humidity.
    Applied Microbiology 23 (6) 1125-1130; 1972
    4

264 Brastad, W.A.; Borchardt, L.F.
Electric hygrometer of small dimensions.
Review of Scientific Instruments 24 (12) 1143-1144; 1953
3
265 Braun, J.V.; Braun, J.D.
A simplified method for preparing solutions of glycerol and
water for humidity control.
Corrosion 14 (3) 117t-118t; 1958
3
266 Brausse, G.; Mayer, A.; Nedetzka, T.; Schlecht, P.; Vogel, H.
Water adsorption and dielectric properties of lyophilized
hemoglobin.
Journal of Physical Chemistry 72 (9) 3098-3105; 1968
3,5
267 Breaden, P.W.; Willoft, E.M.A.
Bread Staling.
Part III. Measurement of the redistribution of moisture in
bread by gravimetry.
Journal of the Science of Food and Agriculture 22 (-)
647-649; 1971
2
268 Breen, S.; Monaghan, R.
Moisture measurement.
In: Gaffney, J.J. (Ed.). Quality Detection in Foods. pp 102-105.
American Society of Agricultural Engineers; 1976
3
269 Breese, M.H.
Hysteresis in the hygroscopic equilibria of rough rice at 25°C.
Cereal Chemistry 32 (11) 481-487; 1955
2,3,5
270 Bremecker, K.D.; List, P.H.
Das Differential-Hygro-Calorimeter (DHC) - ein Gerät zur
Untersuchung der Hygroskopizität in strömender Luft.
Chemie-Ingenieur-Technik 52 (7) 592-594; 1980
2,3
271 Breuer, M.M.; Kennerley, M.G.
The hydration of synthetic polypeptides.
Journal of Colloid and Interface Science 37 (1) 124-131; 1971
2
272 Brey, W.S.jr.; Evans, T.; Hitzrot, L.
Nuclear magnetic relaxation times of water sorbed by proteins.
Lysozyme and serum albumin.
Journal of Colloid and Interface Science 26 (3) 306-316; 1968
2
273 Brey, W.S. jr.; Heeb, M.A.; Ward, T.M.
Dielectric measurements of water sorbed on ovalbumin and
lysozyme.
Journal of Colloid and Interface Science 30 (1) 13-20; 1969
2,5
274 Brian, P.L.T.
Predicting activity coefficients from liquid phase solubility
limits.
Industrial and Engineering Chemistry Fundamentals
4 (1) 100-101; 1965
3
275 Briggs, D.R.
Water relationships in Colloids. II. Bound water in colloids.
Journal of Physical Chemistry 36 (-) 367-386; 1932
2

276 Brockington, S.F.; Dorin, A.C.; Howerton, H.K.
     Hygroscopic equilibria of whole kernel corn.
     Cereal Chemistry 26 (-) 166-173; 1949
     3,5
277 Brockmann, M.C.
     Development of intermediate moisture foods for military use.
     Food Technology 24 (8) 896-900; 1970
     4,5
278 Brockmann, M.C.
     Intermediate moisture foods.
     In: The Skylab Program. USA, Research Development Associates
     for Military Food & Packaging Systems Inc. Activities Report
     25 (1) 70-77; 1973
     1,4
279 Brockmann, M.C.
     Intermediate moisture foods.
     In: Brockmann, M.C.: Food Dehydration, 2nd edition. Vol. 2,
     Practices and Applications. AVI Publ. Comp. Inc. Westport,
     Conn.; 1973
     4
280 Brockmann, R.; Acker, L.
     Verhalten der Lipoxygenase in wasserarmem Milieu.
     I. Einfluß der Wasseraktivität auf die enzymatische
     Lipidoxidation.
     Lebensmittel-Wissenschaft und -Technologie 10 (1) 24-27; 1977
     4
281 Brockmann, R.; Acker, L.
     Wasseraktivität und enzymatische Reaktionen.
     Lebensmittel Technologie 11 (4) 2-7; 1978
     1,4
282 Brodersen, R.; Hangaard, B.J.; Jakobsen, C.; Pedersen, A.O.
     Main chain sorption of water by serum albumin.
     Acta Chemica Scandinavica 27 (-) 573-581; 1973
     2
283 Bromley, L.A.
     Thermodynamic properties of strong electrolytes in aqueous
     solutions.
     American Institute of Chemical Engineers, Journal
     19 (2) 313-320; 1973
     2,3
284 Brooks, J.
     Dried eggs. III. The relation between water content and
     chemical changes during storage.
     Journal of the Society of Chemical Industry 62 (-) 137-139;
     1943
     4
285 Browne, C.A.
     Moisture absorptive power of different sugars and carbohydrates
     under varying conditions of atmospheric humidity.
     Journal of Industrial and Engineering Chemistry
     14 (8) 712-714; 1922
     2,5
286 Brügel, D.; Verbeek, A.E.
     Neue Methoden der titrimetrischen Wasserbestimmung.
     Labor Praxis 4 (1, 2) 18-26; 1980
     3
287 Brunauer, S.
     The adsorption of gases and vapors. Vol. 1, Physical
     adsorption.
     Princeton University Press, London; 1943
     1

288 Brunauer, S.; Deming, L.S.; Deming, W.E.; Teller, E.
    On a theory of the van der Waals adsorption of gases.
    Journal of the American Chemical Society 62 (7) 1723-1732;
    1940
    1
289 Brunauer, S.; Emmet, P.H.; Teller, E.
    Adsorption of gases in multimolecular layers.
    Journal of the American Chemical Society 60 (2) 309-319; 1938
    2
290 Brunauer, S.; Skalny, J.; Bodor, E.
    Adsorption on nonporous solids.
    Journal of Colloid and Interface Science 30 (4) 546-552; 1969
    2
291 Bryan, W.P.
    The thermodynamics of water protein interactions.
    Journal of Theoretical Biology 87 (-) 639-661; 1980
    1,2
292 Bulatov, D.S.
    Determination of the isotherm of sorption of water by proteins
    using infrared spectroscopy.
    Biophysics T 19 (-) 448-453; 1974
    3
293 Buleon, A.; Bizot, H.; Delage, M.M.; Multon, J.L.
    Evolution of crystallinity and specific gravity of potato
    starch versus water ad- and desorption.
    Starch/Stärke 34 (11) 361-366; 1982
    2,4,5
294 Bull, H.B.
    Adsorption of water vapor by proteins.
    Journal of the American Chemical Society 66 (-) 1499-1507;
    1944
    2,5
295 Bull, H.B.; Breese, K.
    Protein hydration. 1. Bindig sites.
    Archives of Biochemistry and Biophysics 128 (2) 488-496; 1968
    2
296 Bull, H.B.; Breese, K.
    Protein hydration. 2. Specific heat of egg albumin.
    Archives of Biochemistry and Biophysics 128 (2) 497-502; 1968
    2
297 Bull, H.B.; Breese, K.
    Water and solute binding by proteins.
    Archives of Biochemistry and Biophysics 137 (-) 299-305; 1970
    2
298 Buma, T.J.
    The physical structure of milk powder and changes which take
    place during moisture adsorption.
    Netherlands Milk Dairy Journal 20 (2) 91-97; 1966
    2,4
299 Bungartz, H.
    Die künstliche Getreidetrocknung.
    VDI-Zeitschrift 97 (-) 364; 1955
    5
300 Burrage, L.J.
    Studies on adsorption. Part VII. The form of isothermals of
    vapor on charcoal and its relation to hysteresis.
    Transaction of the Faraday Society 30 (-) 317; 1934
    2

301 Burrows, I.E.; Barker, D.
Intermediate moisture petfoods.
In: Davies, R.; Birch, G.G.; Parker, K.J. (Edts.).
Intermediate Moisture Foods. pp 43-53.
Applied Science Publ., London; 1976
3,4,5

302 Burvall, A.; Asp, N.G.; Bosson, A.; San José, C.; Dahlqvist,A.
Storage of lactose-hydrolysed dried milk:
Effect of water activity on the protein nutritional value.
Journal of Dairy Research 45 (-) 381-389; 1978
4

303 Bushill, J.H.; Wright, W.B.; Fuller, C.H.F.; Bell, A.V.
The crystallisation of lactose with particular reference to
its occurence in milk powder.
Journal of the Science of Food and Agriculture 16 (-) 622-628;
1965
2,3

304 Bushuk, W.; Hlynka, I.
Weight and volume changes in wheat during sorption and
desorption of moisture.
Cereal Chemistry 37 (-) 390-398; 1960
2,4,5

305 Bushuk, W.; Winkler, C.A.
Sorption of water vapor on wheat flour, starch and gluten.
Cereal Chemistry 34 (2) 73-86; 1957
2,3,5

306 Bushuk, W.; Winkler, C.A.
Sorption of organic vapors on wheat flour at 27 °C.
Cereal Chemistry 34 (3) 87-93; 1957
2,5

307 Busk, G.C. Jr.
Water binding properties of macromolecular systems.
Ph.D. Thesis, University of Minnesota, Minneapolis, USA; 1979
2

308 Bustrillos, A.D.; Banzon, J.
The equilibrium moisture of copra at various relative
humidities.
Philipine Agriculturist 33 (2) 77-87; 1949
Cited in: Gough, M.C.; Lippiat, G.A.
Moisture Humidity Equilibria of Tropical Stored Produce.
Part II-Oilseeds.
Tropical Stored Products Information 34 (-) 49-61; 1977
5

309 Butcher, J.
Moisture determination.
Australasian Baker and Millers' Journal 75 (5) 35-38; 1972
3

310 Buxton, P.A.
The measurement and control of atmospheric humidity in
relation to entomological problems.
Bulletin of Entomological Research 22 (-) 431-447; 1931
4

311 Buxton, P.A.; Mellanby, K.
The measurement and control of humidity.
Bulletin of Entomological Research 25 (-) 171-175; 1934
3

312 Cabral, A.C.D.; Alvim, D.D.
Alimentos desidratados - Conceitos basicos para sua embalagem
e conservacao.
Boletim do Instituto de Tecnologia de alimentos 18 (1) 1-65;
1981
1

313 Cabral, A.C.D.; Fernandes, M.H.C.
    Embalagem para café torrado e café torrado e moido.
    Boletim do Instituto de Tecnologia de alimentos 19 (1) 1-19;
    1982
    4,5

314 Cakebread, S.H.
    Chemistry of candy.
    II. Determining equilibrium relative humidity.
    The Manufacturing Confectioner 49 (3) 29-31; 1969
    3

315 Cakebread, S.H.
    How to control moisture migration in composite product.
    Candy and Snack Industry 137 (4) 24-70; 1972
    5

316 Cakebread, S.H.
    Confectionary ingredients. Vapor pressures of carbohydrate
    solutions.
    Confectionary Production 38 (8) 407-410; 1972
    3

317 Cakebread, S.H.
    Confectionary ingredients. Vapor pressures of carbohydrate
    solutions.
      I. Confectionary Production 38 (8) 407-410; 1972
     II. Confectionary Production 38 (9) 486-496; 1972
    III. Confectionary Production 38 (10) 524-550; 1972
     IV. Confectionary Production 38 (11) 585-604; 1972
      V. Confectionary Production 38 (12) 638-668; 1972
    2,3

318 Cakebread, S.H.
    Confectionary ingredients. Osmotic properties of carbohydrate
    solutions.
    VIII. Confectionary Production 39 (10) 532-537; 1973
      IX. Confectionary Production 39 (11) 588-593; 1973
       X. Confectionary Production 39 (12) 634-636; 1973
    2,3

319 Cakebread, S.H.
    Confectionary ingredients. Osmotic properties of carbohydrate
    solutions - XII.
    Confectionary Production 40 (2) 67-78; 1974
    2,3

320 Cakebread, S.H.
    Confectionary ingredients. Osmotic properties of carbohydrate
    solutions - XIII.
    Confectionary Production 40 (3) 104-109; 1974
    2,3

321 Cal-Vidal, J.
    Agua e preservacao de alimentos.
    Bol. SBCTA/NURESC 5(-) 3-10; 1978
    2,4

322 Cal-Vidal, J.
    O efeito da combinacao de solutos na depressao da atividade
    d'agua.
    Cienc. Tecnol. Aliment. 1 (2) 73-84; 1981
    2

323 Cal-Vidal, J.
    Potencial higroscopico como indice de estabilidade de graos e
    cereais desidratados.
    Pesquisa Agropecuaria Brasileira 17 (1) 61-76; 1982
    2

324 Cal-Vidal, J.
Comportamento higroscópico e poder autoaglomerante de suco de maracuja liofilizado.
Thesis, Universidade de Sao Paulo, Brazil; 1982
2,4,5

325 Cal-Vidal, J.; De Gois, V.A.
Kinetics of water vapor sorption by freeze dried papaya.
Proc. of III. International Symposium and Dryex,
Birmingham, England; 1982
2,5

326 Camposortega, C.; Moncada, F.; Rowen, J.W.
Is the adsorption of water vapor by wool photosensitive?
Journal of the Society of Chemical Industry 68 (-) 118-119;
1949
4

327 Campanini, M.
Sviluppo e sopraovivenza di salmonelle e stafilococchi in relazione all' attività dell' acqua.
Industria Conserve 53 (-) 270-276; 1978
4

328 Campanini, M.; Casolari, A.
Effetto della temperatura e dell' attività dell' acqua sull' accrescimento dei lattobacilli.
Industria Conserve 55 (-) 103; 1980
4

329 Canella, M.; Castriotta, G.; Ilario, L.d'
Viscosita e assorbimento di umidita di farine,concentrati ed isolati proteici di girasole e di soia.
Rivista Italiana delle Sostanze Grasse 54 (9) 394-397; 1977
5

330 Cano Munoz, G.; Hermida Bun, J.R.; Contreras Cano, J.
Humedades de equilibrio de orujos de aceitunas.
Influenca del contenido en humedad y del tiempo de almucenamiento en la calidad de los aceites extraidos con hexano.
Grasas y Aceites 27 (4) 245-253; 1976
5

331 Cardew, M.H.; Eley, D.D.
The sorption of water by haemoglobin.
In: Fundamental Aspects of the Dehydration of Foodstuffs.
pp 24-30. Society of Chemical Industry, London; 1958
2,5

332 Careri, G.; Giansanti, A.; Gratton, E.
Lysozyme film hydration events: an IR and gravimetric study.
Biopolymers 18 (-) 1187-1203; 1979
2,4

333 Carr, A.R.; Townsend, R.E.; Badger, W.L.
Vapor pressures of glycerol-water and glycerol-water-sodium chloride systems.
Industrial and Engineering Chemistry 17 (7) 643-646; 1925
3

334 Carr, D.S.; Harris, B.L.
Solutions for maintaining constant relative humidity.
Industrial and Engineering Chemistry 41 (9) 2014-2015; 1949
3

335 Carter, E.P.; Young, G.Y.
Effect of moisture content, temperature and length of storage on the development of sick wheat in sealed containers.
Cereal Chemistry 22 (-) 418-428; 1945
4

336 Casolari, A.; Bertoli, P.; Castelvetri, F.
Studio preliminare delle caratteristiche termiche di batteri
mesofili.
Industria Conserve 56 (2) 92-96; 1981
4

337 Casolari, A.; Spotti, E.; Castelvetri, F.
Attività dell' acqua e velocita di accrescimento microbico.
Industria Conserve 53 (3) 168-173; 1978
4

338 Casoli, U.; Cultrera, R.
Ricerche sulla conservazione della polvere di pomodoro
liofilizzata.
Industria Conserve 42 (-) 3-7; 1967
5

339 Cassie, A.B.D.
Multimolecular adsorption.
Transactions of the Faraday Society 41 (-) 450-458; 1945
1,2

340 Cassie, A.B.D.
Absorption of water by wool.
Transactions of the Faraday Society 41 (-) 458-464; 1945
2

341 Cassie, A.B.D.
Multimolecular absorption II.
Transactions of the Faraday Society 43 (-) 615; 1947
2

342 Caurie, M.
A new model equation for predicting safe storage moisture
levels for optimum stability of dehydrated foods.
Journal of Food Technology 5 (3) 301-307; 1970
2,5

343 Caurie, M.
A single layer moisture absorption theory as a basis for the
stability and availability of moisture in dehydrated foods.
Journal of Food Technology 6 (2) 193-201; 1971
2

344 Caurie, M.
A practical approach to water sorption isotherms and the basis
for the determination of optimum moisture levels of dehydrated
foods.
Journal of Food Technology 6 (1) 85-93; 1971
2

345 Caurie, M.
Derivation of full range moisture sorption isotherms.
In: Rockland, L.B.; Stewart, G.F. (Edts.).
Water Activity: Influences on Foodquality pp 63-87.
Academic Press, New York; 1981
2

346 Caurie, M.
A monolayer equation for use in sorption studies.
Journal of Food Science 47 (1) 332-333; 1982
2

347 Caurie, M.
Raoult's law, water activity and moisture availability in
solutions.
Journal of Food Science 48 (2) 648-649; 1983
2

348 Caurie, M.; Tung-Ching Lee; Salomon, M.; Chichester, C.O.
A rearranged B.E.T. plot for a more direct estimation of
B.E.T. constants.
Journal of Food Science 41 (2) 448; 1976
2

349 Cavaleru, A.; Comsa, G.; Iosifescu, B.
Extraction for information about activation energy distribution
from desorption studies.
Supplemento al Nuovo Cimento 5 (2) 549-557; 1967
2

350 Cerny, G.
Möglichkeiten zur Verlängerung der mikrobiellen Stabilität von
Lebensmitteln mit verringerter Gleichgewichtsfeuchtigkeit,
dargestellt am Modell Sandkuchen.
Chemie, Mikrobiologie, Technologie der Lebensmittel
5 (1) 20-26; 1976
4

351 Cerofolini, G.F.; Cerofolini, M.
Heterogeneity, allostericity and hysteresis in adsorption of
water by proteins.
Journal of Colloid and Interface Science 78 (1) 65-73; 1980
2

352 Challa, G.
Water in polymers.
Plastica 22 (5) 204-208 1969
Plastica 22 (6) 250-253; 1969
2

353 Chau, K.V.; Heinis, J.J.; Perez, M.
Sorption isotherms and drying rates for mullet fillet and roe.
Journal of Food Science 47 (6) 1318-1322, 1328; 1982
5

354 Chaudet, J.H.; Laine, N.R.; Burton, J.S.
Interrelationships between storage stability and moisture
sorption properties of dehydrated foods.
U.S. Army Natick Laboratories, Final Report Contract No DA
19-129-AMC 252 (N)
2,4

355 Cheftel, C.; Bousquet-Ricard, M.; Quayle, G.; Guilbert, S.;
Elguezabal, L.; Masset, R.
Aliments a humidité intermédiaire: gels polysaccharidiques et
protéiques.
Annales de la Nutrition et de l'Alimentation 32 (-) 597-615;
1978
4,5

356 Chen, C.C.
Water activity in intermediate moisture foods as a function of
water binding, solute activity, and surface tension.
Ph.D. Thesis, Rutgers Univ., New Brunswick, New Jersey, USA;
1977
1,2,4

357 Chen, C.C.; Karmas, E.
Effect of surface active agents on water activity in
intermediate moisture foods.
Lebensmittel-Wissenschaft und -Technologie 12 (2) 68-71; 1979
4

358 Chen, C.C.; Karmas, E.
Solute activity effect on water activity.
Lebensmittel-Wissenschaft und -Technologie 14 (2) 101-104;
1980
2,3

359 Chen, C.S.
Equilibrium moisture curves for biological materials.
American Society of Agricultural Engineers, Paper No.
69-889, Dec.; 1969
2,5

360 Chen, C.S.
Equilibrium moisture curves for biological materials.
Transactions of the American Society of Agricultural Engineers
14 (-) 924-926; 1971
2,5
361 Chen, C.S.; Clayton, J.T.
Equilibrium moisture curves for biological materials - The
effect of temperature.
American Society of Agricultural Engineers Paper No. 70-383;
1970
2,5
362 Chen, C.S.; Clayton, J.T.
The effect of temperature on sorption isotherms of biological
materials.
Transactions of the American Society of Agricultural Engineers
14(5) 927-929; 1971
2,5
363 Chen, D.H.T.; Thompson, A.R.
Isobaric vapor-liquid equilibria for the systems
glycerol-water and glycerol- water saturated with sodium
chloride.
Journal of Chemical and Engineering Data 15 (4) 471-474; 1970
3
364 Chen, H.; Sangster, J.; Teng, T.T.; Lenzi, F.
A general method of predicting the water activity of ternary
aqueous solutions from binary data.
Canadian Journal of Chemical Engineering 51 (4) 234-241; 1973
2,3
365 Childers, A.B.; Terrell, R.N.; Craig, T.M.; Kayfus, T.J.;
Smith, G.C.
Effect of sodium chloride concentration, water activity,
fermentation method and drying time on the viability of
Trichinella spiralis in Genoa Salami.
Journal of Food Protection 45 (9) 816-819; 1982
4
366 Chilton, W.C.; Collison, R.
Hydration and gelation of modified potato starches.
Journal of Food Technology 9 (1) 87-93; 1974
2,5
367 Chirife, J.
Prediction of water activity in intermediate moisture foods.
Journal of Food Technology 13 (5) 417-424; 1978
2,4,5
368 Chirife, J.
A survey of existing sorption data.
In: Jowitt, R.; Escher, F.; Hallström, B.: Meffert, H.F.T.;
Spieß, W.E.L.; Vos, G. (Edts.). Physical Properties of Foods.
pp 55-62.
Applied Science Publ., London, New York; 1983
1
369 Chirife, J.; Alzamora, S.M.; Ferro Fontan, C.
Microbial growth at reduced water activities: studies of aw
prediction in solutions of compatible solutes.
Journal of Applied Bacteriology 54 (3) 339-343; 1983
2,3,4
370 Chirife, J.; Boquet, R.; Iglesias, H.A.
The mathematical description of water sorption isotherm of
foods in the high range of water activity.
Lebensmittel-Wissenschaft und -Technologie 12 (3) 150-152;
1979
2

371   Chirife, J.; Favetto, G.; Ferro-Fontan, C.
      The water activity of fructose solutions in the intermediate
      moisture range.
      Lebensmittel-Wissenschaft und -Technologie 15 (3) 159-160; 1982
      2,3,4
372   Chirife, J.; Favetto, G.; Ferro-Fontan, C.; Resnik, S.L.
      The water activity of standard saturated salt solutions in the
      range of intermediate moisture foods.
      Lebensmittel-Wissenschaft und -Technologie 16 (1) 36-38; 1983
      3
373   Chirife, J.; Favetto, G.; Scorza, O.C.
      The water activity of common liquid bacteriological media.
      Journal of Applied Bacteriology 53 (-) 219-222; 1982
      4,5
374   Chirife, J.; Ferro-Fontan, C.
      Considerations regarding the water activity of aqueous
      solutions in connection with the formulation of intermediate
      moisture foods.
      Proc. of 2nd. International Congress on Engineering and Food,
      pp 337-341. Helsinki; 1979
      2,4,5
375   Chirife, J.; Ferro-Fontan, C.
      Prediction of water activity of aqueous solutions in
      connection with intermediate moisture foods: 5. Experimental
      investigation of the aw lowering behaviour of sodium lactate
      and some related compounds.
      Journal of Food Science 45 (4) 802-804; 1980
      1,4
376   Chirife, J.; Ferro-Fontan, C.
      Water activity of aqueous lactulose solutions.
      Journal of Food Science 45 (6) 1706-1707; 1980
      4,5
377   Chirife, J.; Ferro-Fontan, C.
      A study of the water activity lowering behaviour of poly-
      ethylene glycols in the intermediate moisture range.
      Journal of Food Science 45 (6) 1717-1719; 1980
      3,4
378   Chirife, J.; Ferro-Fontan, C.
      Water activity of fresh foods.
      Journal of Food Science 47 (2) 661-663; 1982
      5
379   Chirife, J.; Ferro-Fontan, C.; Benmergui, E.A.
      The prediction of water activity in aqueous solutions in
      connection with intermediate moisture foods.
      IV. aw-prediction in aqueous nonelectrolyte solutions.
      Journal of Food Technology 15 (1) 59-70; 1980
      4
380   Chirife, J.; Ferro-Fontan, C.; Boquet, R.
      Correlation of water activity data in whole tomato
      concentrates.
      Journal of Food Science 46 (3) 947, 949; 1981
      2,5
381   Chirife, J.; Ferro-Fontan, C.; Scorza, O.C.
      A study of the water activity lowering behaviour of some amino
      acids.
      Journal of Food Technology 15 (4) 383-387; 1980
      4
382   Chirife, J.; Ferro-Fontan, C.; Vigo, M.S.
      A study of water activity prediction for molasses solutions.
      Journal of Agriculture and Food Chemistry 29 (5) 1085-1086;
      1981
      2,5

383 Chirife, J.; Iglesias, H.A.
Equations for fitting water sorption isotherms of foods:
I-A review.
Journal of Food Technology 13 (3) 159-174; 1978
1,2

384 Chirife, J.; Iglesias, H.A.; Boquet, R.
Some characteristics of the heat of water vapour sorption in
dried foodstuffs.
Journal of Food Technology 12 (6) 605-613; 1977
2

385 Chirife, J.; Iglesias, H.A.; Boquet, R.
On the utilization of the B.E.T. equation to estimate the heat
of water sorption in foods.
Lebensmittel-Wissenschaft und -Technolgie 11 (4) 222-223; 1978
2

386 Chirife, J.; Karel, M.
Effect of structure disrupting treatments on volatile release
from freeze-dried maltose.
Journal of Food Technology 9 (1) 13-20; 1974
2,4

387 Chirife, J.; Scorza, O.C.; Vigo, M.S.; Bertoni, M.H.;
Cattaneo, P.
Preliminary studies on the storage stability of intermediate
moisture beef formulated with various water binding agents.
Journal of Food Technology 14 (4) 421-428; 1979
4,5

388 Chirife, J.; Vaamonde, G.; Scarmato, G.
On the minimal water activity for growth of Staphylococcus
aureus.
Journal of Food Science 47 (6) 2054-2057; 1982
4

389 Chirife, J.; Vigo, M.S.; Scorza, O.C.; Bertoni, M.H.;
Cattaneo, P.
Retention of available lysine after long term storage of
intermediate moisture beef formulated with various humectants.
Lebensmittel-Wissenschaft und -Technologie 13 (1) 44-45; 1980
4

390 Chlikadze, A.M.; Aschchiyan, O.A.; Zantaschwili, D.I.
Untersuchungen zur Hygroskopizität von Calciumtartrat (Orig.
russ.)
Vinodelie i Vinogradarstvo SSSR - (5) 15-17; 1978
5

391 Choi, H.Y.; Kim, M.N.; Lee, K.H.
Non-enzymatic browning reactions in dried squid stored at
different water activities.
Bulletin of the Korean Fisheries Society 6 (3/4) 97-100; 1973
4

392 Chomanov, U.CH.; Tumenov, S.N.; Kakimov, A.K.
Equipment for determination of water activity.(Orig.russ.)
Myasnaya Industrya SSSR. 2 (-) 33-34; 1981
3

393 Chorbajian, T.
Moisture equilibria in seeds.
Ph.D. Thesis, Iowa State College; 1958
2,5

394 Chordash, R.A.; Potter, N.N.
Effects of dehydration through the intermediate moisture range
on water activity, microbial growth, and texture of selected
foods.
Journal of Milk and Food Technology 35 (7) 395-398; 1972
4,5

395 Chou, D.H.; Morr, C.V.
     Protein water interactions and functional properties.
     Journal of American Oil Chemist's Society 56 (-) 53A-62A; 1979
     2,4,5
396 Chou, H.E.; Acott, K.M.; Labuza, T.P.
     Sorption hysteresis and chemical reactivity: lipid oxidation.
     Journal of Food Science 38 (2) 316-319; 1973
     2,4,5
397 Chou, H.E.; Breene, W.
     Oxidative decoloration of ß-carotene in low moisture model
     systems.
     Journal of Food Science 37 (1) 66-68; 1972
     4
398 Chou, H.E.; Labuza, T.P.
     Antioxidant effectiveness in intermediate moisture content
     model systems.
     Journal of Food Science 39 (-) 479-483; 1974
     4,5
399 Choudhury, M.S.
     The effects of moisture adsorption on the tensile strength of
     rice.
     Ph.D. Thesis, A and M University, College Station, Texas, USA;
     1970
     3,4
400 Chowdhury, S.
     Germination of fungal spores in relation to atmospheric
     humidity.
     Indian Journal of Agricultural Science 7 (-) 653; 1937
     4
401 Christensen, C.M.
     Lagerschäden durch Pilzbefall bei Getreide.
     Getreide und Mehl 11 (-) 78-81; 1962
     4
402 Christensen, C.M. (Ed.).
     Storage of cereal grains and their products.
     American Association of Cereal Chemists, Incorporated,
     St. Paul, Minnesota; 1974
     2,3,4,5
403 Christensen, C.M.; Kaufmann, H.H.
     Microflora
     In: Christensen, C.M. (Ed.). Storage of Cereal Grains and
     their Products. pp 158-192. American Association of Cereal
     Chemists, Incorporated, St. Paul, Minnesota; 1974
     4
404 Christensen, C.M.; Linko, P.
     Moisture contents of hard red winter wheat as determined by
     meters and by oven drying, and influence of small differences
     in moisture content upon subsequent deterioration of the grain
     in storage.
     Cereal Chemistry 40 (-) 129-137; 1963
     3,4
405 Christian, J.H.B.
     Microbiological spoilage problems of dehydrated foods.
     In: Microbiological Quality of Foods. pp 223-228. Academic Press
     New York; 1963
     4
406 Christian, J.H.B.
     Water activity and the growth of microorganisms.
     In: Leitch, M.J.; Rhodes, D.N. (Edts.). Recent Advances in
     Food Science. Vol. 3, pp 248-255. Butterworth, London; 1963
     4

407 Christian, J.H.B.
Specific solute effects on microbial water relations.
In: Rockland, L.B.; Stewart, G.F. (Edts.).
Water Activity: Influences on Foodquality. pp 825-854.
Academic Press, New York; 1981
4

408 Christian, J.H.B.; Michener, H.D.; Jarvis, B. (Edts.).
Safety of food as influenced by water activity.
Parma, Italy 1978
1,4

409 Christian, J.H.B.; Scott, W.J.
Water relations of salmonellae at 30 °C.
Australian Journal of Biological Sciences 6 (-) 565-573; 1953
4

410 Christian, J.H.B.; Stewart, B.
Survival of Staphylococcus aureus and Salmonella newport in
dried foods, as influenced by water activity and oxygen.
In: Hobbs, B.; Christian, J.(Edts.). The Microbiological Safety
of Food. pp 107-119. Academic Press, London; 1973
4

411 Christian, J.H.B.; Waltho, J.A.
Solute concentrations within cells of halophilic and
non-halophilic bacteria.
Biochimica et Biophysica Acta 65 (-) 506-508; 1962
2,4

412 Christian, J.H.B.; Waltho, J.A.
The water relations of staphylococci and micrococci.
Journal of Applied Bacteriology 25 (3) 369-377; 1962
4

413 Chuang, L.; Toledo, R.T.
Predicting the water activity of multicomponent systems from
water sorption isotherms of individual component.
Journal of Food Science 41 (4) 922-927; 1976
2,5

414 Chung, C.Y.; Toyomizu, M.
Studies on the browning of dehydrated foods as a function of
water activity.
1. Effect of aw on browning in amino acid-lipid systems.
Bulletin of the Japanese Society of Scientific Fisheries
42 (6) 697-702; 1976
4

415 Chung, D.S.
Thermodynamic factors influencing moisture equilibrium of
cereal grains and their products.
Ph.D. Thesis, Kansas State University; 1966
2,3,5

416 Chung, D.S.; Hodges, T.; Pfost, H.B.
Hygroscopic properties of soybean and sesame seed.
Annual meeting of the American Association of Cereal Chemists
59th, Montreal 1974
Abstract in: Cereal Science Today 19 (9) 402; 1974
5

417 Chung, D.S.; Park, S.W.; Hoover, W.J.; Watson, C.A.
Sorption kinetics of water vapor by yellow dent corn.
II. Analysis of kinetic data for damaged corn.
Cereal Chemistry 49 (5) 598-604; 1972
2,5

418 Chung, D.S.; Pfost, H.B.
     Adsorption and desorption of water vapor by cereal grains and
     their products.
     Part 1: Heat and free energy changes of adsorption and
     desorption.
     Transactions of the American Society for Agricultural
     Engineers 10 (4) 549-551, 555; 1967
     2,5
419 Chung, D.S.; Pfost, H.B.
     Adsorption and desorption of water vapor by cereal grains and
     their products.
     Part 2: Development of general isotherm equation.
     Transactions of the American Society for Agricultural
     Engineers 10 (4) 552-555; 1967
     2,5
420 Chung, D.S.; Pfost, H.B.
     Adsorption and desorption of water vapor by cereal grains and
     their products.
     III. A hypothesis for explaining the hysteresis effect.
     Transactions of the American Society for Agricultural
     Engineers 10 (4) 556-557; 1967
     2,5
421 Chyr, C.Y.L.; Walker, H.W.; Hinz, P.
     Influence of pH, temperature, curing agents, and water
     activity on germination of PA 3679 spores.
     Journal of Food Protection 40 (6) 369-372; 1977
     4
422 Ciner-Doruk, M.; Eichner, K.
     Bildung und Stabilität von Amadori-Verbindungen in wasserarmen
     Lebensmitteln.
     Zeitschrift für Lebensmittel-Untersuchung und Forschung
     168 (1) 9-20; 1979
     4.
423 Clayton, C.N.
     The germination of fungous spores in relation to controlled
     humidity.
     Phytopathology 32 (-) 921; 1942
     4
424 Cleland, J.E.; Fetzer, W.R.
     Moisture absorptive power of starch hydrolysates.
     Industrial and Engineering Chemistry 36 (6) 552-555; 1944
     4,5
425 Cliver, D.; Kostenbader, K.; Vallenas, M.
     Stability of viruses in low moisture foods.
     Journal of Milk and Food Technology 33 (11) 484-491; 1970
     4
426 Coelho, U.; Miltz, J.; Gilbert, S.G.
     Application of inverse phase gas chromatography for
     determination of bound water in collagen.
     Journal of Food Science 44 (4) 1150-1151; 1979
     2,5
427 Cohan, L.H.
     Sorption hysteresis and the vapor pressure of concave
     surfaces.
     Journal of the American Chemical Society 60 (-) 433; 1938
     2
428 Cohan, L.H.
     Hysteresis and the capillary theory of adsorption of vapors.
     Journal of the American Chemical Society 66 (-) 98-105; 1944
     2

429   Cohen, E.; Saguy, I.
      Effect of water activity and moisture content on the stability
      of beet powder pigments.
      Journal of Food Science 48 (3) 703-707; 1983
      4,5
430   Cointot, A.; Cruchaudet, J.; Simonot-Grange, M.H.
      Mise en oeuvre d'une méthode thermogravimétrique automatique
      pour le trace des isostères d'adsorption.
      Comptes rendues de l'Academie des Sciences Paris; Serie C,
      269 (4) 302-305; 1969
      2,3
431   Colas, B.
      Changes in the water content and in the water activity of dry
      sausages during maturation, studied in three experiments.
      Alimentation et la Vie 64 (2/3) 125-130; 1976
      4,5
432   Coleman, D.A.; Fellows, H.C.
      Hygroscopic moisture of cereals, grains and flax seed exposed
      to atmospheres of different relative humidity.
      Cereal Chemistry 2 (9) 275-287; 1925
      5
433   Coleman, D.A.; Rothgeb, B.E.; Fellows, H.C.
      Respiration of sorghum grains.
      Technical Bulletin United States Department of Agriculture
      100 (-) 1-16; 1928
      Cited in: Gough,M.C.; Bateman,G.A.
      Tropical Stored Products Information 33 (-) 25-40; 1977
      5
434   Collins, E.M.
      The partial pressures of water in equilibrium with aqueous
      solutions of sulfuric acid.
      Journal of Physical Chemistry 37 (-) 1191-1203; 1933
      3
435   Collins, J.L.; Chen,; C.C.; Park, J.R.; Mundt, J.O.;
      McCarty, I.E.; Johnston, M.R.
      Preliminary studies on some properties of intermediate
      moisture, deep-fried fish flesh.
      Journal of Food Science 37 (-) 189; 1972
      4
436   Collins, J.L.; Yu, A.K.
      Stability and acceptance of intermediate moisture, deep-fried
      cat fish.
      Journal of Food Science 40 (-) 858-863; 1975
      4
437   Collison, R.; Chilton, W.
      Starch gelation as a function of water content.
      Journal of Food Technology 9 (3) 309-315; 1974
      2,4
438   Colvin, R.; Craig, G.M.; Sallans, H.R.
      Hygroscopic equilibria for hulls and kernels of sunflower seed
      and oats.
      Canadian Journal of Research 25F (2) 111-118; 1947
      3,5
439   Cooke, R.; Kuntz, I.D.
      The properties of water in biological systems.
      Annual Review of Biophysics and Bioengineering 3 (-) 95-126;
      1974
      1

440   Cooper, D.L.
      Studies of salt fish. III. Equilibrium moisture coefficients
      of salt fish muscle.
      Journal of the Fisheries Research Board of Canada
      4 (2) 136-140; 1938
      5
441   Cooper, D.L.; Noel, T.C.
      Some equilibrium moisture values of fresh and salt cod.
      Journal of the Fisheries Research Board of Canada 23 (5)
      775-778; 1966
      5
442   Cooper, R.M.; Knight, R.A.; Robb, J.; Seiler, D.A.L.
      Equilibrium relative humidity of cakes.
      Food Trade Review 38 (4) 40-45, 54; 1968
      3,4,5
443   Cope, F.W.
      A theory of cell hydration governed by adsorption of water on
      cell proteins rather than by osmotic pressure.
      Bulletin of Mathematical Biophysics 29 (-) 583-596; 1967
      2
444   Coppens, B.; Wei, H.
      La détermination du point hygroscopique et de la courbe
      d'hygroscopicité en fonction de la teneur en eau dans les
      produits agricoles à sécher.
      Annales de Gembloux 60 (4) 225-235; 1954
      1,3
445   Corey, H.
      Texture in foodstuffs.
      Critical Reviews in Food Technology 5 (-) 161-198; 1970
      4,5
446   Corry, J.E.L.
      The water relations and heat resistance of microorganisms.
      Scientific and Technical Surveys of British Food Manufacturing
      Industries Research Association 73; 1971
      4,5
447   Corry, J.E.L.
      The water relations and heat resistance of micro-organisms.
      Progress in Industrial Microbiology 12 (-) 73-108; 1973
      4
448   Corry, J.E.L.
      The effect of water activity on the heat resistance of
      bacteria.
      In: Duckworth, R.B. (Ed.).
      Water Relations of Foods. pp 325-337. Academic Press,
      London; 1975
      4
449   Corry, J.E.L.
      The safety of intermediate moisture foods with respect to
      salmonella.
      In: Davies, R.; Birch, G.G.; Parker, K.J. (Edts.).
      Intermediate Moisture Foods. pp 215-238. Applied Science
      Publ., London; 1976
      4
450   Corry, J.E.L.
      Relationships on water activity to fungal growth.
      In: Beuchat, L.R. (Ed.). pp 45-82.
      Food and Beverage Mycology. AVI Publ. Comp. Inc. Westport,
      Conn.; 1978
      4,5

451 Corte Dos Santos, A.; Hahn, D.; Cahagnier, B.; Drapron, R.;
     Guilbot, A.; Lefebvre, J.; Multon, J.L.; Poisson, J.;
     Trentesaux, E.
     Etude de l'évolution de plusieurs charactéristiques d'un café
     arabica au cours d'un stockage expérimental effectué à cinq
     humidités relatives différentes.
     Café-Cacao-Thé 15 (4) 329-340; 1971
     3,4,5
452 Cramer, W.D.; Hanson, T.P.; Tsao, G.T.; Lancaster, E.B.
     Water vapor sorption in starch granules.
     Journal of Macromolecular Science - Phys. 3 (4) 611-622; 1969
     2,3
453 Crapiste, G.H.; Rotstein, E.
     Prediction of sorptional equilibrium data for starch-containing
     foodstuffs.
     Journal of Food Science 47 (5) 1501-1507; 1982
     2,5
454 Crawford, H.G.
     Partial pressure of water over aqueous solutions of sulfuric
     acid.
     Industrial Engineering Chemistry 17 (5) 522-523; 1925
     3
455 Crothers, D.M.; Ratner, D.I.
     Thermodynamic studies of a model system for hydrophobic
     bonding.
     Biochemistry 7 (5) 1823-1827; 1968
     2
456 Cuendet, L.S.; Larson, E.; Norris, C.A.; Geddes, W.F.
     The influence of moisture content and other factors on the
     stability of wheat flours at 37,8°C.
     Cereal Chemistry 31 (10) 362-389; 1954
     4
457 Culpepper, C.W.; Caldwell, J.S.; Wright, R.C.
     Effect of temperature and atmospheric humidity upon the
     behaviour of dehydrated white potatoes in storage.
     The Canner 104 (14) 14, 16, 18-20; 1947
                104 (15) 15-16, 18, 26; 1947
                104 (16) 16, 18, 20, 22, 24, 30; 1947
                104 (17) 27-28, 30, 32; 1947
     4
458 Cunniff, L.C.
     Newer techniques for process measurement.
     Food Product Development 11 (2) 76-78; 1977
     3
459 Curran, H.R.
     Influence of osmotic pressure upon spore germination.
     Journal of Bacteriology 21 (3) 197-209; 1931
     4
460 Cutting, C.L.; Reay, G.A.; Sherwan, J.M.
     Dehydration of fish.
     Special report No. 62, Department of Scientific and Industrial
     Research, London 1956
     5
461 Daiber-Kuhnke, U.
     Das Feuchtigkeitsgleichgewicht von Luft und Getreide bei der
     Behältertrocknung.
     Landtechnische Forschung 9 (4) 106-111; 1955
     2,5
462 Dall'Aglio, G.; Gherardi, S.; Balestrazzi, A.
     Stabilata enzimatica in rapporto all'umidita relativa.
     Industria Conserve 47 (2) 96-99; 1972
     4

463 Dall'Aglio, G.; Balestrazzi, A.; Gherardi, S.; Versitano, A.
Le attivita enzimatiche in rapporto all'umidita relativa.
Industria Conserve 47 (1) 35-41; 1972
4,5

464 Dall'Aglio, G.; Porretta, A.; Versitano, A.
Impiego del riscaldamento a microonde a piastre ed a raggi
infrarossi nella liofilizzazione, dei gamberetti et dei
calamari.
Industria Conserve 46 (2) 93-100; 1971
5

465 Dallyn, H.; Fox, A.
Spoilage of materials of reduced water activity by xerophilic
fungi.
Society of Applied Bacteriology.
Techn. Ser. 15 (-) 129-139; 1980
4

466 Damköhler, G.
Über die Adsorptionsgeschwindigkeit von Gasen an porösen
Adsorbentien.
Zeitschrift für physikalische Chemie A 174 (-) 222-238; 1953
2

467 Daniels, N.W.R.
Some effects of water in wheat flour doughs.
In: Duckworth, R.B. (Ed.).
Water Relations of Foods. pp 573-586. Academic Press, London;
1975
5

468 Danilatos, G.; Feughelman, M.
The internal dynamic mechanical loss in alpha-keratin fibers
during moisture sorption.
Textile Research Journal 46 (11) 845-846; 1976
4

469 Danilatos, G.D.; Postle, R.
Dynamic mechanical properties of keratin fibers during water
absorption and desorption.
Journal of Applied Polymer Science 26 (-) 193-200; 1981
4

470 D'Arcy, R.L.; Watt, I.C.
The uptake of sulfuric acid vapor by wool.
Journal of the Textile Institute 57 T (-) 416-424; 1966
2,3

471 D'Arcy, R.L.; Watt, I.C.
Analysis of sorption isotherm of non homogeneous sorbents.
Transactions of the Faraday Society 66 (-) 1236-1245; 1970
2,5

472 D'Arcy, R.L.; Watt, I.C.
Water vapor sorption isotherms on macromolecular substrates.
In: Rockland, L.B.; Stewart, G.F. (Edts.).
Water Activity: Influences on Foodquality. pp 111-142.
Academic Press, New York; 1981
2,5

473 Das, B.
Sorption-desorption of water vapor by wheat flours.
Deutsche Lebensmittel-Rundschau 70 (4) 139-141; 1974
5

474 Das, B.; Sethi, R.K.; Chopra, S.L.
Sorption and desorption of water vapor on starch.
Israel Journal of Chemistry 10 (-) 963-965; 1972
2,5

475 Davey, F.K.
Hair humidity elements.
In: Wexler, A.; Ruskin, R.E. (Edts.).
Humidity and Moisture. Measurement and Control in Science and
Industry. Vol. 1, pp 571-573. Reinhold Publ. Corporation,
New York; 1965
3

476 Davey, P.M.; Elcoate, S.
Moisture content / relative humidity equilibrium of tropical
stored produce.
Tropical Stored Products Information
Part   I: (11) 439-467; 1965
Part   II: (12) 495-512; 1966
Part III: (13) 15-34; 1967
5

477 Davies, R.; Birch, G.G.; Parker, K.J. (Edts.).
Intermediate Moisture Foods.
Applied Science Publ., London; 1976
1,2,3,4,5

478 Davis, D.S.
Vapour pressure nomographs for aqueous sodium hydroxide
solutions.
The Journal of Industrial and Engineering Chemistry
34 (-) 1131-1132; 1942
3

479 Davis, E.G.
Evaluation and selection of flexible films for food packaging.
Food Technology in Australia 22(2) 62-67; 1970
2,5

480 Davis, S.; McLaren, A.D.
Free energy, heat and entropy changes accompanying the
sorption of water vapor by proteins.
Journal of Polymer Science 3 (1) 16-21; 1948
2

481 Dawson, P.T.
The transition between localized and condensed layers in the
adsorption of water vapor onto titania.
Journal of Physical Chemistry 71 (4) 838-844; 1967
2

482 Day, D.L.; Nelson, G.L.
Desorption isotherms for wheat.
Transaction of the American Society of Agricultural Engineers
8 (-) 293-297; 1965
2,5

483 Dean, E.W.; Stark, D.D.
A convenient method for the determination of water in
petroleum and other organic emulsions.
Journal of Industrial and Engineering Chemistry
12 (5) 486-490; 1920
3

484 De Boer, J.H.
The dynamical character of adsorption.
Clarendon Press, Oxford; 1953
1,2

485 De Boer, J.H.; Zwikker, C.
Adsorption als Folge von Polarisation;
die Adsorptionsisotherme.
Annalen der Physik und Chemie B 3 (-) 407-418; 1929
2

486 De Gois, V.A.
Comportamento higroscópico do mamao liofilizado com vistas ao
estabelicamento do seu potencial de "caking".
Thesis, University Lavras - MG, Brasil; 1981
2,3,4,5
487 De Gois, V.A.; Cal-Vidal, J.
Termodinamica da sorcao e desorcao de mamao liofilizado em pó
e em grânulos.
Proc. of III. Encontro Nacional de Secagem. Vicosa, M.G.
Brazil; 1981
2,5
488 Delaney, R.A.M.
Protein concentrates from slaughter animal blood.
II. Composition and properties of spray dried red blood cell
concentrates.
Journal of Food Technology 12 (4) 355-368; 1977
5
489 Delmer, F.
Les produits à humidité intermédiaire dans l'industrie
alimentaire.
Industries Alimentaires et Agricoles 93 (9/10) 1141-1148; 1976
1,4,5
490 Demeyer, D.
Vergleich zwischen berechneten und gemessenen Werten der
Wasseraktivität (aw) in Rohwurst.
Fleischwirtschaft 59 (7) 940-943; 1979
1,3,5
491 Demirdzic, I.; Kaludjercic, P.
Über den Einfluß des Lewis-Faktors auf die Genauigkeit der
Feuchtemessung mit Psychrometer.
Technisches Messen atm - (7/8) 283-285; 1978
3
492 Deng, J.J.
Protein-protein interaction and its role in fat and water
binding in comminuted flesh products.
Ph.D. Thesis, University of Georgia, USA; 1974
2
493 Dengler, W.
Praktische Bedeutung von Wasserdampf-Sorptionsisothermen.
Chemie-Anlagen und -Verfahren 4 (-) 68-75; 1971
1,2,3,5
494 Dengler, W.; Blenke, H.
Untersuchung des Einflusses der Alterung technischer
Adsorbentien auf das Adsorptionsgleichgewicht und die
Sorptionsgeschwindigkeit.
Chemie-Ingenieur-Technik 46(6) 244-245; 1975
1,2
495 Denizel, T.; Rolfe, E.J.; Jarvis, B.J.
Moisture-equilibrium relative humidity relationships in
pistachio nuts with particular regard to control of aflatoxin
formation.
Journal of the Science of Food and Agriculture
27 (-) 1027-1034; 1976
3,4,5
496 Dennison, D.; Kirk, J.; Bach, J.; Kokoczka, P.; Heldman, D.
Storage stability of thiamin and riboflavin in a dehydrated
food system.
Journal of Food Processing and Preservation 1 (1) 43-54; 1977
4

497 Dent, R.W.
     A multilayer theory for gas sorption.
     Part I: Sorption of a single gas.
     Journal of Textile Research 47 (2) 145-152; 1977
     2
498 Derby, R.; Miller, B.; Miller, F.; Trimbo, H.
     Visual observation of wheat-starch gelatinisation in limited
     water systems.
     Cereal Chemistry 52 (5) 702-713; 1975
     2
499 Desrosier, N.W.
     The technology of food preservation, 3. ed. Chapter 11:
     preservation of semi-moist foods. pp 365-383.
     AVI Publ. Comp. Inc. Westport, Conn.; 1970
     4
500 De Vor, H.
     Een onderzoek naar de reproduceerbaarheid van aw-waarden
     gemeten met behulp van een Lufft-aw-Wert-Messer, model 5803.
     Voedingsmiddelentechnologie 9 (36) 10-11; 1976
     3,5
501 Dexter, S.T.; Anderson, A.L.; Pfahler, P.L.; Benne, E.J.
     Responses of white pea beans to various humidities and
     temperatures of storage.
     Agronomy Journal 47 (6) 246-250; 1955
     4,5
502 Diaz-Santanilla, J.; Fritsch, G.; Müller-Warmuth, W.
     NMR-Untersuchungen über innere Bewegungen und über den Einbau
     von Wasser in Roh- und Röstkaffee.
     Zeitschrift für Lebensmittel Untersuchung und -Forschung
     172 (-) 173-177; 1981
     2
503 Dickens, J.W.; Pattee, H.E.
     The effect of time, temperature and moisture on aflatoxin
     production in peanuts inoculated with a toxic strain of
     Aspergillus flavus.
     Tropical Science 8 (-) 11-22; 1966
     4
504 Diener, L.U.; Davis, N.D.
     Limiting temperature and relative humdity for growth and
     production of aflatoxin and free fatty acids by Aspergillus
     flavus in sterile peanuts.
     The Journal of the American Oil Chemists Society
     44 (4) 259-263; 1967
     4
505 Dierchen, W.
     Über das Verderben von Waffelfüllungen durch den
     Mikrobengehalt von Rohstoffen.
     Brot und Gebäck 7 (9) 140-143; 1953
     4
506 Dingle, J.R.
     Moisture curves for light salted cod.
     Fisheries Research Board of Canada. Prog. Rep. Atlantic Coast
     Sta. No. 53 (-) 8-9; 1952
     5
507 Dittmar, J.H.
     Hygroscopicity of sugars and sugar mixtures.
     Journal of Industrial and Engineering Chemistry
     27 (3) 333-335; 1935
     5

508 Doe, P.E.; Hashmi, R.; Poulter, R.G.; Olley, J.
     Isohalic sorption isotherms
     I. Determination for dried salted cod (Gadus morrhua).
     Journal of Food Technology 17 (1) 125-134; 1982
     2,5
509 Doe, P.E.; Poulter, R.G.; Curran, C.A.
     Determination of water activity and shelf-life of salted dried
     fish from moisture and salt content measurements.
     Agricultural Organization Fish Technology in Africa.
     Symposium 268 (Suppl.) 113-120; 1982
     2
510 Doebbler, G.F.
     Mechanisms of freezing injury to biological systems.
     Final Report.
     Union Carbide Corporation, Linde Division, Tonawanda Research
     Laboratory; Contract Nr. 4788(00); 1966
     5
511 Doelle, H.J.; Riekert, L.
     Kinetik der Sorption, Desorption und Diffusion in Zeolithen.
     Angewandte Chemie 91 (4) 309-316; 1979
     2,3
512 Dole, M.
     Statistical thermodynamics of the sorption of vapors by solids
     Journal of Chemical Physics 16 (1) 25-30; 1948
     2
513 Dole, M.; McLaren, A.D.
     The free energy, heat and entropy of sorption of water vapor
     by proteins and high polymers.
     Journal of the American Chemical Society 69 (3) 651-657; 1947
     2
514 Dollimore, D.; Holt, B.
     Sorption of vapors in cellulose film.
     Journal of Applied Polymer Science 17 (-) 1795-1803; 1973
     2
515 Donnelly, B.J.; Fruin, J.C.; Scallet, B.L.
     Reactions of oligosaccharides. III. Hygroscopic properties.
     Cereal Chemistry 50 (4) 512-519; 1973
     4,5
516 Doroszewicz, S.
     Auswertung der Sorption von Wasserdampf im Bohnenkaffee.
     Lebensmittel-Industrie 25 (7) 316; 1978
     2
517 Doull, K.M.; Mew, P.
     The hygroscopic properties of different dilutions of honey.
     Apidologie 8 (1) 19-24; 1977
     5
518 Downes, J.G.; Mackay, B.H.
     Sorption kinetics of water vapour in wool fibres.
     Journal of Polymer Science 28 (1) 45-67; 1958
     2,5
519 Dradon, J.
     Nouvelle application du point froid pour le séchage du
     houblon.
     Le Petit J. Brass. 69 (-) 248-251, 267-270; cited in:
     Brauwissenschaft 15 (2) 44; 1962
     3,5
520 Drapron, R.
     Réactions enzymatiques en milieu peu hydraté
     Annales de Technologie Agricole 21 (4) 487-499; 1972
     4

521 Drapron, R.; Guilbot, A.;
Contribution à l'étude des réactions enzymatiques dans les
milieux biologiques peu hydratés: la dégradation de l'amidon
par les amylases en fonction de l'activité de l'eau et de la
température.
Annales de Technologie Agricole 11 (3) 175-218; 1962
                            11 (4) 275-371; 1962
4,5
522 Draudt, H.N.; Damon, C.E.; Huang, I.Y.; Rowe, C.
Enzyme activity in freeze dried foods.
Purdue University Progress Report No. 3; 1960
4
523 Draudt, H.N.; Huang, I.Y.
Effect of moisture content of freeze dried peaches and bananas
on changes during storage related to oxidative and
carbonyl-amine browning.
Journal of Agricultural and Food Chemistry 14 (-) 170-176;
1966
4
524 Dreher, H.; Kast, W.
Analytische Näherungslösung für die Durchbruchskurve bei
nichtlinearer Adsorptionsisotherme.
Chemie-Ingenieur-Technik 51 (2) 122-124; 1979
1,2
525 Dreher, H.; Kast, W.
Gleichgewicht bei der Adsorption mehrerer Komponenten aus der
Gasphase.
Chemie-Ingenieur-Technik 51 (12) 1245; 1979
1,2
526 Dreier, W.; Jakobson, D.
Shelf stable, intermediate moisture, flake textured doughs and
method for making same.
U.S. Patent Nr. 3.769.034; 1973
4
527 Drexler, H.; Gierschner, K.
Apparatur zur automatischen Aufnahme von Sorptionsisothermen.
Chemie-Ingenieur-Technik 43 (11) 691; 1971
3
528 Drexler, H.; Gierschner, K.
Untersuchungen zur Technologie und Qualitätsbeurteilung
zerstäubungsgetrockneter Fruchtpulver.
Industrielle Obst- und Gemüseverwertung 58 (2) 29-36, (3)
57-64, (4) 93-98; 1973
3,4,5
529 Dubinin, M.M.
Fundamentals of the theory of physical adsorption of gases and
vapors in micropores.
In: Ricca,F. (Ed.). Adsorption-Desorption Phenomena.
Academic Press, New York; 1972
1,2
530 Duck, W.; Cross, R.P.
Vapor pressure study of hard candy.
The Manufacturing Confectioner 37 (8) 17-22; 1957
3
531 Duckworth, R.B.
Diffusion of solutes in dehydrated vegetables.
In: Hawthorn, J.; Leitch, M.J. (Edts.).
Recent Advances in Food Science. Vol. 2, pp 46-49. Butterworth,
London; 1962
2

532 Duckworth, R.B.
     The properties of water around the surfaces of food colloids.
     Proc. of the Institute of Food Science and Technology
     (U.K.) 5 (2) 60-67; 1972
     1,2,5
533 Duckworth, R.B.
     Bound water in intermediate moisture foods.
     Proc. Intermediate moisture foods. Seminar 5, CPCIA,
     Paris; 1975
     2,4,5
534 Duckworth, R.B.
     Water Relations of Foods.
     Academic Press, London, New York, San Francisco; 1975
     1,2,3,4,5
535 Duckworth, R.B.
     Roles of water in food.
     Journal of the Science of Food and Agriculture 27 (7) 709-712;
     1976
     1,2
536 Duckworth, R.B.
     Factors influenced by water in foods.
     Chemistry and Industry - (24) 1039-1042; 1976
     1
537 Duckworth, R.B.
     Solute mobility in relation to water content and water
     activity.
     In: Rockland, L.B.; Stewart, G.F. (Edts.).
     Water Activity: Influences on Foodquality. pp 295-317.
     Academic Press, New York; 1981
     4,5
538 Duckworth, R.B.
     Future needs in water sorption in foodstuffs.
     In: Jowitt, R.; Escher, F.; Hallström, B.; Meffert, H.F.T.;
     Spieß, W.E.L.; Vos, G. (Edts.).
     Physical Properties of Foods. pp 93-101.
     Applied Science Publ., London, New York; 1983
     1
539 Duckworth, R.B.; Allison, J.Y.; Clapperton, H.A.A.
     The aqueous environment for chemical change in intermediate
     moisture foods.
     In: Davies, R.; Birch, G.G.; Parker, K.J. (Edts.).
     Intermediate Moisture Foods. pp 89-99.
     Applied Science Publ., London; 1976
     4
540 Duckworth, R.B.; Smith, G.M.
     The environment for chemical change in dried and frozen foods.
     Proc. of the Nutrition Society 22 (-) 182-189; 1963
     4,5
541 Duckworth, R.B.; Smith, G.M.
     Diffusion of solutes at low moisture levels.
     In: Leitch, M.J.; Rhodes, D.N. (Edts.).
     Recent Advances in Food Science Vol. 3, pp 230-238.
     Biochemistry and Biophysics in Food Research.
     Butterworth London; 1963
     2
542 Duden, R.
     Zum Problem enzymatischer Reaktionen in Lebensmitteln bei sehr
     niedriger Wasseraktivität.
     Lebensmittel-Wissenschaft und -Technologie 4 (6) 205-206; 1971
     4

543 Duden, R.
Über die in gefriergetrockneten Champignons, Erbsen und Bohnen
enthaltenen Lipide und deren enzymatische Veränderungen bei
der Lagerung.
Dissertation, Universität Karlsruhe; 1972
4,5

544 Duman, J.G.; Patterson, J.L.; Kozak, J.J.; De Vries, A.L.
Isopiestic determination of water binding by fish antifreeze
glycoproteins.
Biochimica et Biophysica Acta 626 (-) 332-336; 1980
2,3

545 Dunford, H.B.; Morrison, J.L.
The adsorption of water vapor by proteins.
Canadian Journal of Chemistry 32 (-) 558-560; 1954
2

546 Dunford, H.B.; Morrison, J.L.
The heat of wetting of silk fibroin by water.
Canadian Journal of Chemistry 33 (-) 904-912; 1955
2

547 Dunker, C.F.; Tressler, D.K.; Wruch, M.; Blake, K.B.
Relationship of moisture content to quality retention in
dehydrated vegetables during storage.
I. Tomato flake.
Food Technology 1 (11) 17-25; 1947
4

548 Dunstan, E.R.; Chung, D.S.; Hodges, T.O.
Adsorption and desorption characteristics of grain sorghum.
Transactions of the American Society of Agricultural Engineers
16 (4) 667-670; 1973
2,5

549 Duprat, F.; Guilbot, A.
Solvent versus non-solvent water in
starch-alcohol-water-systems.
In: Duckworth, R.B. (Ed.).
Water Relations of Foods. pp 173-182.
Academic Press, London; 1975
2,5

550 Dupuis, R.
Intermediate moisture foodstuffs containing reduced
polysaccharides.
U.S. Patent No. 3.516.838; 1970
4

551 Dushchenko, V.P.; Panchenko, M.S.; D'yachenko, S.F.
Temperature dependance of water-vapor adsorption by
capillary-porous bodies.(Orig.russ.)
Inzhenerno-Fizicheskii Zhurnal 16 (1) 67-71; 1969
2

552 Dyer, D.F.; Carpenter, D.K.; Sunderland, J.E.
Equilibrium vapor pressure of frozen bovine muscle.
Journal of Food Science 31 (-) 196-201; 1966
2,5

553 Dymsza, H.A.; Silverman, G.
Improving the acceptability of intermediate moisture fish.
Food Technology 33 (10) 52-53; 1979
4

554 Dzhevizov, S.; Dimitrova, N.; Kiseva, R.
The aw value of some raw dry sausages produced in Bulgaria.
Proc. of the European Meeting of Meat Research Workers
No. 23, L6:1-L6:19; 1977
5

555 Eagland, D.
Protein-hydration - its role in stabilizing the helix
conformation of the protein.
In: Duckworth, R.B. (Ed.).
Water Relations of Foods. pp 73-92.
Academic Press, London; 1975
2

556 Eberius, E.
Wasserbestimmung mit Karl-Fischer-Lösung.
Angewandte Chemie 66 (5) 121-122; 1954
3

557 Edgar, G.; Swan, W.O.
The factors determining the hygroscopic properties of soluble
substances.
I. The vapor pressure of saturated solutions.
Journal of the American Chemical Society 44 (-) 570; 1922
3

558 Ehler, K.F.; Bernhard, R.A.; Nickerson, T.A.
Heats of adsorption of small molecules on various forms of
lactose, sucrose and glucose.
Journal of Agricultural and Food Chemistry 27 (5) 921-927;
1979
2

559 Ehrlich, S.H.; Bettelheim, F.A.
Infrared spectroscopy of the water sorption process of
mucopolysaccharides.
Journal of Physical Chemistry 67 (-) 1954-1960; 1963
2

560 Eichner,K.,
Nicht enzymatische Veränderungen von Lebensmitteln mit
niedrigen und mittleren Wassergehalten.
36. Diskussionstagung, Forschungskreis der Ernährungsindustrie
e.V. Hannover. pp 19-32; 1977
4

561 Eichner, K.
The influence of water content on non-enzymic browning
reactions in dehydrated foods and model systems and the
inhibition of fat oxidation by browning intermediates.
In: Duckworth,R.B.(Ed.). Water Relations of Foods. pp 417-434.
Academic Press, London; 1975
4

562 Eichner, K.
Chemisch analytischer Nachweis beginnender Qualitätsverände-
rungen bei wasserarmen Lebensmitteln.
Zeitschrift für Lebensmittel-Technologie und Verfahrenstechnik
31 (3) 89-92; 1980
4

563 Eichner, K.; Ciner-Doruk, M.
Bildung und Stabilität von Amadori-Verbindungen in wasserarmen
Modellsystemen.
Zeitschrift für Lebensmittel-Untersuchung und -Forschung
168 (5) 360-367; 1979
4

564 Eichner, K.; Ciner-Doruk, M.
Formation and decomposition of browning intermediates and
visible sugar-amino browning reactions.
In: Rockland, L.B.; Stewart, G.F. (Edts.).
Water Activity: Influences on Foodquality. pp 567-603.
Acedemic Press, New York; 1981
4

565 Eichner, K.; Karel, M.
The influence of water content and water activity on the
sugar-amino browning reaction in model systems under various
conditions.
Journal of Agricultural and Food Chemistry 20 (2) 218-223;
1972
4
566 Ekberg, A.; Fields, M.L.; Edmondson, J.
Effect of storage on lysine-enriched corn, millet and sorghum.
Journal of Food Science 44 (2) 630-631; 1979
4
567 Ekedahl, E.; Silien, L.
A mathematical model for sorption hysteresis. I.
Acta Chemica Scandinavia 19 (10) 2323-2335; 1965
2
568 Elder, L.W.
Fundamentals of package function.
Modern Packaging - (10) 138-142, 196-198; 1949
2,5
569 Elenkov, V.; Gegov, J.
Sorption characteristics examination of some freeze dried
foods of plant origin. - Sorption isotherms.(Orig.bulg.)
Nauchni Trudove, Nauchnoizsledovatelski Institut po Konserva
promishlemost, Plovdiv 13 (-) 89-99; 1977
2,5
570 Eley, D.D.; Leslie, R.B.
Adsorption of water on solid proteins with special reference
to hemoglobin.
Advances in Chemical Physics 7 (-) 238-258; 1963
2,5
571 Eliasson, A.C.
Effect of water content on the gelatinization of wheat starch.
Die Stärke 32 (8) 270-272; 1980
4
572 Elrick, D.E.; Laryea, K.B.
Sorption of water in soils: a comparison of techniques for
solving the diffusion equation.
Soil Science 128 (6) 369-373; 1979
2
573 Elvanides, S.N.
An analysis and interpretation of thermodynamic properties of
foods as determined from moisture sorption data.
Ph.D. Thesis, Michigan State University, USA; 1972
1,2,3,5
574 Elworthy, P.H.; George, T.M.
Sorption of water vapor and surface activity of ghatti gum.
Journal of Pharmacy and Pharmacology 16 (-) 258-264; 1964
2,5
575 Emig, G.; Heilmann, W.
Registrierendes Quecksilber-Manometer zur Aufnahme von
Sorptionsisothermen.
Chemie-Ingenieur-Technik 40 (16) 817-818; 1968
3
576 Enderby, J.A.
Water absorption by polymers.
Transactions of the Faraday Society 51 (-) 106-116; 1955
2
577 Engelson, M.; Solberg, M.; Karmas, E.
Antimycotic properties of hop extract in reduced water
activity media.
Journal of Food Science 45 (5) 1175-1178; 1980
4

578 Erickson, L.E.
Recent developments in intermediate moisture foods.
Journal of Food Protection 45 (5) 484-491; 1982
4,5

579 Erlander, S.R.
The structure of water and its relationship to
hydrocarbon-water interactions.
Journal of Macromolecular Science-Chemistry A 2 (3) 595-621;
1968
2

580 Erlandson, J.; Wrolstad, R.
Degradation of anthocyanins at limited water concentration.
Journal of Food Science 37 (4) 592-595; 1972
4

581 Ernst, E.
Bound water in physics and biology.
Acta Biochimica et Biophysica Academiae Scientarium Hungaricae
5 (1) 57-69; 1970
1,2

582 Eschbach, H.L.
Messung von Adsorptionsisothermen mit Schwingquarzen.
Transactions of the 3rd International Vacuum Congress in
Stuttgart 2 (2) 443-446; 1965
3

583 Eschmann, K.H.
Getrocknete Lebensmittel.
Archiv für Lebensmittelhygiene 6 (-) 126-131; 1970
1,4

584 Esteban-Quilez, M.A.; Marcos-Barrado, A.
Data de sorcion de humedad de la musculatura de las aves con
valores pH finales altos y bajos.
Anales de Bromatologia 26 (2) 149-158; 1974
4,5

585 Esteban-Quilez, M.A.; Marcos-Barrado, A.; Sinha, B.K.
Water activity of honey and the growth of osmotolerant yeasts.
Anales de Bromatologia 28 (1) 33-44; 1976
4

586 Everett, D.H.
A general approach to hysteresis. Part 3. A formal treatment
of the independant domain model of hysteresis.
Transactions of the Faraday Society 50 (-) 1077-1096; 1954
2

587 Everett, D.H.; Whitton, W.I.
A general approach to hysteresis.
Transactions of the Faraday Society 48 (-) 749-757; 1952
2

588 Everett, D.H.; Whitton, W.I.
A thermodynamic study of the adsorption of benzene vapour by
active charcoals.
Proc. of the Royal Society A 230 (-) 91-110; 1955
2,3

589 Fairbrother, T.H.
The influence of environment on the moisture content of flour
and wheat.
Cereal Chemistry 6 (-) 379-395; 1929
5

590 Falk, M.; Hartmann, K.A. jr.; Lord, R.C.
Hydration of deoxyribonucleic acid.
I. A gravimetric study.
Journal of the American Chemical Society 84 (-) 3844-3846;
1962
2,5

591 Fan, L.T.; Chung, P.; Shellenberger, J.A.
Diffusion coefficients of water in wheat kernels.
Cereal Chemistry 38 (11) 540-548; 1961
2

592 Farrow, F.D.; Swan, E.
The absorption of water by dried films of boiled starch.
Journal of Textile Institute, Transactions 14 (-) T 465-T 474;
1923
3,5

593 Favetto, G.; Chirife, J.; Bartholomai, G.B.
Determination of moisture content in glycerol-containing
intermediate moisture foods.
Journal of Food Science 44(4) 1258-1259; 1979
3,4

594 Favetto, G.; Chirife, J.; Bartholomai, G.B.
A study of water activity lowering in meat during immersion
cooking in sodium chloride-glycerol solutions.
I. Equilibrium considerations and diffusional analysis of
solute uptake.
Journal of Food Technology 16 (6) 609-619; 1981
2,4

595 Favetto, G.; Chirife, J.; Bartholomai, G.B.
A study of water activity lowering in meat during
immersion-cooking in sodium chloride-glycerol solution.
II. Kinetics of aw lowering and effect of some process
variables.
Journal of Food Technology 16 (6) 621-628; 1981
2,4

596 Favetto, G.; Resnik, S.; Chirife, J.; Ferro-Fontan, C.
Statistical evaluation of water activity measurements obtained
with the Vaisala humicap humidity meter.
Journal of Food Science 48 (2) 534-538; 1983
3

597 Fedors, R.F.
Osmotic effects in water absorption by polymers.
Polymer 21 (-) 207-212; 1980
2

598 Feldman, R.F.; Sereda, P.J.
A model for hydrated Portland cement paste as deduced from
sorption-length change and mechanical properties.
Materiaux et Constructions 1 (6) 509-520; 1968
1,2

599 Felt, C.F.; Cook, F.H.; Borchardt, L.F.
The determination of a satisfactory type of packaging for the
bulk shipment of dehydrated apple slices by freight.
Journal of Food Technology 6 (-) 390-393; 1952
4,5

600 Fennema, O.R.
Water and protein hydration.
In: Whitaker, J.R.; Tannenbaum,S.R. (Edts.). Food Proteins.
pp 50-90. AVI Publ. Comp. Inc. Westport, Conn.; 1977
2

601 Fennema, O.R.
Enzyme kinetics at low temperature and reduced water
activity.
In: Crowe, J.H.; Clegg, J.S. (Edts.).
Dried Biological Systems. pp 297-322.
Academic Press, New York; 1978
2,4,5

602 Fennema, O.R.
     Water activity at subfreezing temperatures.
     In: Rockland, L.B.; Stewart, G.F. (Edts.).
     Water Activity: Influences on Foodquality. pp 713-732.
     Academic Press, New York; 1981
     1,5
603 Fennema, O.R.; Berny, L.A.
     Equilibrium vapor pressure and water activity of food at
     subfreezing temperatures.
     Proc. of the IV. International Congress of Food Science
     and Technology, Madrid Vol. 2, 27-35; 1974
     3,5
604 Fenton, F.C.
     Storage of grain sorghum.
     Agricultural Engineering, St. Joseph Michigan, 22 (-) 185-188;
     1941
     4,5
605 Fernandez-Salguerro Carretero, J.
     Introduccion al estudio de las propriedades de sorcion de
     humedad de los alimentos semsecos: bacalao salado y seco.
     Archivos de Zootecnia 22 (87) 241-270; 1973
     1,5
606 Ferrell, R.E.; Shepherd, A.D.; Thielking, R.; Pence, J.W.
     Moisture equilibrium of bulgur.
     Cereal Chemistry 43 (1) 136-142; 1966
     5
607 Ferro-Fontan, C.; Chirife, J.
     The evaluation of water activity in aqueous solutions from
     freezing point depression.
     Journal of Food Technology 16 (1) 21-30; 1981
     2,5
608 Ferro-Fontan, C.; Chirife, J.
     A refinement of Ross's equation for predicting the water
     activity of non-electrolyte mixtures.
     Journal of Food Technology 16 (2) 219-221; 1981
     2,4
609 Ferro-Fontan, C.; Benmergui, E.A.; Chirife, J.
     The prediction of water activity of aqueous solutions in
     connection with intermediate moisture foods.
     III. aw-prediction in multicomponent strong electrolyte
     aqueous solutions.
     Journal of Food Technology 15 (1) 47-58; 1980
     4
610 Ferro-Fontan, C.; Chirife, J.; Benmergui, E.A.
     The prediction of water activity in aqueous solutions in
     connection with intermediate moisture foods.
     II. On the choice of the best aw-lowering single strong
     electrolyte.
     Journal of Food Technology 14 (6) 639-646; 1979
     4
611 Ferro-Fontan, C.; Chirife, J.; Boquet, R.
     Water activity in multicomponent non-electrolyte solutions.
     Journal of Food Technology 18 (5) 553-559; 1981
     2,3
612 Ferro-Fontan, C.; Chirife, J.; Sancho, E.; Iglesias, H.A.
     Analysis of a model for water sorption phenomena in foods.
     Journal of Food Science 47 (5) 1590-1594; 1982
     2,3
613 Fetkenheuer, W.
     Bedeutung der relativen Luftfeuchtigkeit für den Masseverlust
     bei der Lagerung von Kernobst.
     Gartenbau 25 (10) 308-309; 1978
     4

614 Fett, H.M.
Water activity determination in foods in the range 0.80
to 0.99.
Journal of Food Science 38 (6) 1097-1098; 1973
3,5
615 Fidler, F.
Packaging of freeze-dried foods.
Food Trade Review 34 (5) 40-44; 1964
4,5
616 Fidler, J.C.F.; Gane, R.; Davies, R.; Mapson, L.W.; Pinder,
J.L.; Bishop, E.A.
Dried soup powders.
I. Precooked.
Food Manufacture 20 (-) 277-281; 1945
4
617 Filonenko, G.K.; Tschuprin, A.I.
Gleichgewichtsfeuchte von Lebensmitteln.(Orig.russ.)
Inzhenerno-Fizicheskii Zhurnal 8 (1) 98-104; 1967
5
618 Fine, M.S.; Olson, A.G.
Tallowiness or rancidity in grain products.
Industrial Engineering Chemistry 20 (6) 652-654; 1928
4
619 Finn-Kelsey, P.; Hulbert, D.G.
The relationship between relative humidity and the moisture
content of agricultural products.
Technological Report of British Electrical Research
Association W/T 33; 1957
cited in: Gough,M.C.;Batemann,G.A.; Tropical Stored Products
Information 33(-) 25-40; 1977
5
620 Fischer, A.; Basel, H.; Kotter, L.
Zur lebensmittelrechtlichen Beurteilung von Rohwurst nach dem
aw-Wert.
Die Fleischwirtschaft 56 (2) 184-187; 1976
4
621 Fish, B.P.
Diffusion and equilibrium properties of water in starch.
Food Investigation Technical Paper, Her Majesty's
Stationery Office (H.M.S.O.), London; 1957
2,5
622 Fish, B.P.
Diffusion and thermodynamics of water in potato starch gel.
In: Fundamental Aspects of Dehydration of Foodstuffs.
pp 143-157. Society of Chemical Industry, London; 1958
5
623 Fisher, D.A.; Jacobson, R.L.; Pflug, I.J.
Indices of water content in gaseous systems, their
measurement, and relationship to each other.
Journal of Milk and Food Technology 38 (11) 706-714; 1975
1,3,4
624 Fisher, E.A.; Jones, C.R.
A note on moisture interchange in mixed wheats, with
observations on the rate of absorption of moisture by wheat.
Cereal Chemistry 16 (9) 573-583; 1939
2
625 Fisher, L.R.; Israelachvili, J.N.
Direct experimental verification of the Kelvin equation for
capillary condensation.
Nature 277 (5697) 548-549; 1979
2

626 Fito, P.J.; Sanz, F.J.
Equilibrium moisture contents in drying of rice bran.
Proc. IV. International Congress of Food Science and Technology.
Madrid. Vol. 4, pp 477-488; 1974
2,3,5

627 Flink, J.M.
The retention of volatile components during freeze drying: a
structurally based mechanism.
In: Goldblith, S.A.; Rey, L.; Rothmayr, W.W. (Edts.).
Freeze Drying and Advanced Technology. pp 351-372.
Academic Press, London; 1975
2,5

628 Flink, J.M.
Intermediate moisture food products in the american market
place.
Journal of Food Processing and Preservation 1 (4) 324-339;
1977
4

629 Flink, J.M.
Considerations regarding the formulation of intermediate
moisture foods.
CPCIA Course on Intermediate Moisture Foods, Nov. 1977, Paris
4

630 Flink, J.; Karel, M.
Mechanisms of retention of organic volatiles in freeze dried
systems.
Journal of Food Technology 7 (2) 199-211; 1972
2,5

631 Flora, L.F.; Beuchat, L.R.; Rao, V.N.M.
Preparation of a shelf-stable, intermediate moisture food
product from muscadine grape skins.
Journal of Food Science 44 (-) 854-856; 1979
4

632 Flory, P.J.
Principles of polymer chemistry.
Cornell University Press, Ithaca, London; 1953
2

633 Flückiger, W.; Cleven, F.
Wasseraktivität; ihre Bedeutung für die Haltbarkeit von
Gebäcken.
Backtechnik 26 (6) 13-15; 1978
1,5

634 Fogiel, A.; Heller, W.
Sorption of vapors by proteins.
I. Sorption of water vapor and ethanol vapor by egg albumin.
Journal of Physical Chemistry 70 (6) 2039-2043; 1966
2,5

635 Folman, M.; Yates, D.J.C.
Infra-red and length-change studies in adsorption of H_2O and
CH_3OH on porous silica glass.
Transactions of the Faraday Society 54 (-) 1684-1691; 1958
2

636 Foster, A.G.
Sorption hysteresis. Part I. Some factors determining the
size of the hysteresis loop.
The Journal of Physical and Colloid Chemistry 55 (-) 638-643;
1951
2

637 Foster, A.G.
    Sorption hysteresis. Part II. The role of the cylindrical
    meniscus effect.
    Journal of the Chemical Society, London (-) 1806-1812; 1952
    2
638 Fowler, R.; Guggenheim, E.A.
    Statistical thermodynamics.
    Cambridge University Press, Cambridge; 1939
    1,2
639 Fox, D.C.; Katz, M.J.
    In: Wolsky, S.P.; Zdanuk, E.J. (Edts.).
    Ultra Micro Weight Determination in Controlled Environments.
    Interscience Publ., New York, Sydney, Toronto; 1969
    3
640 Fox, K.K.; Holsinger, V.H.; Harper, M.K.; Howard, N.; Pryor,
    L.S.; Pallansch, M.J.
    The measurement of the surface·areas of milk powders by a
    permeability procedure.
    Food Technology 17 (6) 127-130; 1963
    3
641 Fox, M.
    Untersuchungen zur mikrobiologischen Stabilisierung von
    wasserreichen Lebensmitteln durch kombinierte Verfahren unter
    Berücksichtigung minimaler Schädigung von Inhaltsstoffen.
    Ph.D. Thesis, University Karlsruhe; 1980
    4
642 Fox, M.; Loncin, M.
    Investigations into the microbiological stability of
    water-rich foods processed by a combination of methods.
    Lebensmittel-Wissenschaft und -Technologie 15 (6) 321-325;
    1982
    4
643 Fox, M.; Loncin, M.; Weiss, M.
    Investigations into the influence of water activity, pH and
    heat treatment on the breakdown, of thiamine in foods.
    Journal of Foodquality 5 (-) 161-182; 1982
    4
644 Frank, H.K.; Betancourt, L.
    Die Paranuß; Teil 1: Herkunft, Gewinnung, chemische und
    physikalische Eigenschaften.
    Deutsche Lebensmittel-Rundschau 76 (1) 7-11; 1980
    5
645 Frank, H.K.; Betancourt, L.; Uboldi, M.
    Die Paranuß; Teil 2: Verderb und Bildungsbedingungen für
    Aflatoxine.
    Deutsche Lebensmittel-Rundschau 76 (2) 47-50; 1980
    4
646 Frankenfeld, J.W.; Karel, M.; Labuza, T.P.
    Intermediate moisture food compositions containing aliphatic
    1,3-diols.
    United States Patent 3 732 112; 8th May 1973
    4
647 Franks, F.
    Water, ice and solutions of simple molecules.
    In: Duckworth, R.B. (Ed.).
    Water Relations of Foods. pp 3-22.
    Academic Press, London; 1975
    2
648 Frasco, C.B.
    Water structure and heat denaturation in lysozyme.
    Ph.D. Thesis, Rutgers University, New Brunswick, New Jersey; 1977
    2,4

649 Fraser, C.W.; Haley, W.L.
Factors that influence the rate of absorption of water by wheat.
Cereal Chemistry 9 (-) 45-49; 1932
2,4

650 Freundlich, H.
Colloid and capillary chemistry.
Methuen & Co., London; 1926
2

651 Frey, H.J.; Moore, W.J.
Adsorption of vapors on organic crystals.
I. Adsorption of water vapor on glycine, leucine, diketopiperazine and diglycylglyzine.
Journal of the American Chemical Society 70 (-) 3644-3649; 1948
2,5

652 Fribourg, J.L.; Bercot, J.; Berquet, P.
Emulsions alimentaires a humidité intermédiaire.
Annales de la Nutrition et de l'Alimentation, Paris
32 (-) 617-630; 1978
4,5

653 Fricker, A.
"Intermediate moisture foods" Eine neue Form konservierter Lebenmittel?
In: Harmer, R.; Auerswald, W.; Zöllner, N.; Blanc, B.; Brandstetter, B. (Edts.). Probleme um neue Lebensmittel.
pp 263-276.
Wilhelm Maudrich Verlag Wien, München, Bern; 1978
4

654 Friesen, J.A.
Predicting equilibrium moisture content within the hysteresis loop.
Transactions of the American Society of Agricultural Engineers
17· (2) 339-341; 1974
2

655 Fritsch, G.; Guex, M.
Influence of water absorption on electron spin resonance signals of radiation-induced free radicals in cellulose.
Nestle Research News pp 91-95; 1973
2,3,5

656 Führer, C.; Rebentisch, M.
Untersuchungen der Sorptionshüllen von pharmazeutisch verwendeten Pulvern. 1. Mitteilung: Theoretische Grundlagen.
Pharmazeutische Industrie 37 (3) 187-192; 1975
2,3

657 Fugassi, P.; Ostapchenko, G.
Sorption of polar vapors on swelling gels.
Fuel 38 (-) 271-276; 1959
2,5

658 Fuller, M.E.; Brey, W.S. jr.
Nuclear magnetic resonance study of water sorbed on serum albumin.
Journal of Biological Chemistry 243 (2) 274-280; 1968
2,5

659 Funk, W.A.
Moisture equilibrium.
Modern Packaging 20 (6) 135-138, 162-163; 1947
3,5

660 Gál, S.
Die Methodik der Wasserdampf-Sorptionsmessungen.
Springer Verlag, Berlin, Heidelberg, New York; 1967
1,2,3,5

661 Gál, S.
Die Wasserdampf-Sorptionsisothermen fester Sorbentien.
Chimia 22 (11)409-425; 1968
1,5
662 Gál, S.
Über die Ausdrucksweisen der Konzentration des Wasserdampfes
bei Wasserdampf-Sorptionsmessungen.
Helvetica Chimica Acta 55 (5) 1752-1757; 1972
1
663 Gál, S.
Die praktische Bedeutung von Wasserdampf-Sorptionsmessungen in
der Lebensmittelindustrie.
Alimenta 11 (6) 213-217; 1972
1
664 Gál, S.
Recent advances in techniques for the determination of
sorption isotherms.
In: Duckworth, R.B. (Ed.).
Water Relations of Foods. pp 139-154.
Academic Press, London; 1975
3
665 Gál, S.
Solvent versus non-solvent water in casein-sodium
chloride-water systems.
In: Duckworth, R.B. (Ed.).
Water Relations of Foods. pp 183-191.
Academic Press, London; 1975
2
666 Gál, S.
Bindungsfähigkeit von Casein für Natriumchlorid bei hohen
Wasseraktivitäten.
Die Makromolekulare Chemie - (178) 1535-1544; 1977
2
667 Gál, S.
Die Bestimmung der Wasseraktivität und von
Sorptionsisothermen.
Lebensmitteltechnologie 12 (2) 12-18; 1979
1,3
668 Gál, S.
Recent developments in techniques for obtaining complete
sorption isotherms.
In: Rockland, L.B.; Stewart, G.F. (Edts.).
Water Activity: Influences on Foodquality. pp 89-110.
Academic Press, New York; 1981
3
669 Gál, S.
The need for, and practical applications of, sorption data.
In: Jowitt, R.; Escher, F.; Hallström, B.; Meffert, H.F.T.;
Spieß, W.E.L.; Vos, G. (Edts.).
Physical Properties of Foods. pp 13-24.
Applied Science Publ., London, New York; 1983
1
670 Gál, S.; Bankay, D.
Hydration of sodium chloride bound by casein at medium water
activities.
Journal of Food Science 36 (5) 800-803; 1971
2,5
671 Gál, S.; Tomka, I.; Signer, R.
Zur mathematischen Behandlung der
Wasserdampf-Sorptionsisothermen quellbarer Stoffe.
Chimia 30 (2) 65-68; 1976
2

672 Galambos, J.
A Mezögazdasági termelési folyamatoc gépesitése.
Mezögazdasági Kiado, Budapest; 1972
5

673 Gallaher, G.L.
A method of determining the latent heat of agricultural crops.
Agricultural Engineering 32 (1) 34, 38; 1951
2,5

674 Gamayunov, N.I.; Vasil'eva, L.Y.U.; Koshkin, V.M.
Investigation of the sorption of water on biopolymers.
Biofizika 20 (1) 38-40; 1975
2

675 Gane, R.
The water content of wheats as a function of temperature and humidity.
Journal of the Society of Chemical Industry, Transactions
60 (-) 44-46; 1941
5

676 Gane, R.
Dried meat. III. The water relations of air-dried pre-cooked beef and pork.
Journal of the Society of Chemical Industry; 62 (-) 139-140; 1943
5

677 Gane, R.
Dried egg.
VI. The water relations of dried egg.
Journal of the Society of Chemical Industry 62 (-) 185-187; 1943
5

678 Gane, R.
The water content of the seeds of peas, soyabeans, linseed, grass, onion and carrot as a function of temperature and humidity of atmosphere.
Journal of Agricultural Science 38 (1) 81-83; 1948
5

679 Gane, R.
The effect of temperature, water content and composition of the atmosphere on the viability of carrot, onion, and parsnip seeds in storage.
Journal of Agricultural Science 38 (1) 84-89; 1948
4

680 Gane, R.
The effect of temperature, humidity and atmosphere on the viability of chewing's fescue grass seed in storage.
Journal of Agricultural Science 38 (1) 90-92; 1948
4

681 Gane, R.
The water relations of some dried fruits, vegetables and plant products.
Journal of the Science of Food and Agriculture 1 (2) 42-46; 1950
4,5

682 Gapalakrishna, A.G.; Prabhakar, J.V.
Effect of water activity on autoxidation of raw peanut oil.
Journal of the American Oil Chemists' Society
60 (5) 968-970; 1983
4

683 Garcia Gomez, R.M.; Cano Munoz, G.; Hermida Bun, J.R.
    Estudio de las isotermas de sorcion de aqua en orujos de
    aceituna.
    I. Modelos teoricos clasicos.
    Grasas y Aceites 30 (6) 383-390; 1979
    II. Influencia de la temperatura y efecto sobre el ciclo de
    histeresis.
    Grasas y Aceites 31 (1) 27-31; 1980
    III. Aplicacion a la estabilidad en almacenamiento.
    Grasas y Aceites 31 (1) 33-36; 1980
    2,4,5

684 Garner, W.E.
    The heats of adsorption and the kinetics of adsorption.
    Transaction of the Faraday Society 28 (-) 261-269; 1932
    2

685 Gascoyne, P.R.C.; Pethig, R.
    Experimental and theoretical aspects of hydration isotherms
    for biomolecules.
    Journal of the Chemical Society: Faraday Trans I, 73 (-)
    171-180; 1977
    2,3

686 Gaßmann, B.
    Intermediate moisture foods - eine neue Klasse von
    Lebensmitteln?
    Ernährungsforschung 23 (3) 85-86; 1978
    4

687 Gast, T.; Gebauer, K.P.; Robens, E.
    Quarzwendel-Manometer.
    Technisches Messen 47 (11) 405-407; 1980
    3

688 Gast, T.; Robens, E. (Edts.).
    Progress in vacuum microbalance techniques; Vol. 1,
    Heyden and Son Ltd., London; 1972
    3

689 Gavrilenko, I.V.; Stefanov, L.P.
    Hygroscopic properties of maize germ.(Orig.russ.)
    Trudy Vsesoyuznogo Nauchno-Issledovatel'skogo Instituta Zhirov
    27 (-) 221-225; 1970
    2,5

690 Gay, F.J.
    Effect of temperature on moisture content-equilibrium of
    wheat.
    Journal of the Council for Scientific and Industrial Research,
    Australia 19 (-) 187-189; 1946
    5

691 Gee, M.; Farkas, D.; Rahman, A.R.
    Some concepts for the development of intermediate moisture
    foods.
    Food Technology 31 (4) 58-64; 1977
    4,5

692 Gerber, H.E.
    A saturation hygrometer for the measurement of relative
    humidity between 95 and 105%.
    Journal of Applied Meteorology 19 (10)1196-1208; 1980
    3

693 Gerhardt, U.
    Bevorratung von Gewürzen (II.).
    -Wasserdampf-Sorptionsisothermen von Gewürzen-.
    Die Fleischwirtschaft 54 (7) 1132-1134; 1974
    2,3,5

694 Gerschenson, L.N.; Bartholomai, G.B.; Chirife, J.
Structure collapse and volatile retention during heating and
rehumidification of freeze dried tomato juice.
Journal of Food Science 46 (5) 1552-1556; 1981
4

695 Geurts, T.; Walstra, P.; Mulder, H.
Water binding to milk protein, with particular reference to
cheese.
Nederlands Melk en Zuiveltijdschrift 28 (1) 46-72; 1974
2

696 Giannini, A.A.
Controlled air protects quality in food packaging rooms.
Food Industries 16 (3) 68-72, 135; 1944
4,5

697 Giauque, W.F.; Hornung, E.W.; Kunzler, J.E.; Rubin, T.R.
The thermodynamic properties of aqueous sulfuric acid
solutions and hydrates from 15-300 K.
Journal of the American Chemical Society 82 (-) 62-70; 1960
2,3

698 Gibbard, H.F.; Scatchard, G.; Rousseau, R.A.; Creek, J.L.
Liquid vapor equilibrium of aqueous sodium chloride, from 298
to 373 K and from 1 to 6 mol kg-1, and related properties.
Journal of Chemical and Engineering, Data 19 (3) 281-288; 1974
2,3

699 Gibert, H.; Angelino, H.; Besombes-Vailhé, J.
Généralisation de la theorie thermodynamique de l'adsorption
physique.
Chimie et Industrie - Génie chimique 100 (9) 1424-1430; 1968
2

700 Gibson, B.
The effect of high sugar concentrations on the heat resistance
of vegetative micro-organisms.
Journal of Applied Bacteriology 36 (-) 365-376; 1973
4

701 Giddey, C.
Intermediate moisture foods.
Alimentaria 84 (-) 35, 38-39; 1977
4

702 Gierschner, K.; Drexler, H.
Beitrag zur Beurteilung von Fruchtpulvern aufgrund spezieller
analytischer Kennwerte.
Gordian 69 (8) 371-373, (9) 447-450; 1969
3,5

703 Gildemeister, E.; Räuber, H.J.
Kontinuierliche Feuchtigkeitsmessung an Lebensmitteln über
hygroskopische Sättigung mittels Hochfrequenztechnik.
Lebensmittel-Industrie 22 (3) 111-114; 1975
3

704 Gildemeister, E.; Räuber, H.J.
Kontinuierliche dielektrische Feuchtigkeitsmessungen an
Lebensmitteln.
Lebensmittel-Industrie 23 (-) 489-494; 1976
3

705 Gildemeister, E.; Räuber, H.J.
Kontinuierliche dielektrische Feuchtigkeitsmessungen an
Lebensmitteln.
2. Mitt. Meßgerät zur kontinuierlichen dielektrischen
Feuchtigkeitsmessung.
Lebensmittel-Industrie 25 (-) 23-25; 1978
3

706 Ginzburg, A.S.
The forms and energy of moisture binding in food as a basis
for choosing rational methods for processing and storage.
In: Rockland, L.B.; Stewart, G.F. (Edts.).
Water Activity: Influences on Foodquality. pp 679-711.
Academic Press, New York; 1981
2,4

707 Ginzburg, A.S.; Kostrikov, P.V.; Zozulevich, B.v.;
Karazhiya, V.F.
Equilibrium moisture content of fruits. (Orig. russ.)
Izvestiya Vysshikh Uchebnykh Zavedenii, Pishchevaya
Tekhnologiya - (5) 168-170; 1977
3,5

708 Ginzburg, B.Z.
The binding energy of water to bovine serum albumin.
Journal of Colloid and Interface Science 85 (2) 422-430; 1982
2

709 Glueckauf, E.; Kitt, G.P.
Thermodynamic data on concentrated sulfuric acid solutions.
Transactions of the Faraday Society 52 (-) 1074-1079; 1956
2,3

710 Goepfert, J.M.; Iskander, I.K.; Amundson, C.H.
Relation of the heat resistance of salmonellae to the water
activity of the environment.
Applied Microbiology 19 (3) 429-433; 1970
4

711 Görling, P.
Versuche zur Ermittlung des Trocknungsverhaltens von
Kartoffelstücken.
VDI-Forschungsheft 458, Beilage zu "Forschung auf dem Gebiete
des Ingenieurwesens", Ausgabe B, 22 (-) 5-35; 1956
2,3,5

712 Görling, P.
Einfluß der Lagerbedingungen auf die Qualitätserhaltung von
Trockenerzeugnissen.
I. Untersuchungen über das Lagerverhalten von luftgetrocknetem
Weißkohl, Wirsing, Rotkohl und luftgetrockneten Karotten.
Die industrielle Obst- und Gemüseverwertung 47 (22) 673-676;
1962
4

713 Görling, P.
Über die Qualitätserhaltung von Trockenerzeugnissen.
III. Das Trocknen von Gemüse und Kartoffeln auf niedrigere
Endwassergehalte.
Die Industrielle Obst- und Gemüseverwertung 48 (2) 32-38; 1963
2,5

714 Gokeen, N.A.
Vapor pressure of water above saturated lithium chloride
solution.
Journal of the American Chemical Society 73 (-) 3789; 1951
3

715 Good, R.J.
The contact angle of water on polystyrene: a study of the
cause of hysteresis.
Journal of Colloid and Interface Science 66 (2) 360-362; 1978
2

716 Gore, H.C.; Mangels, C.E.
The relation of moisture content to the deterioration of
raw-dried vegetables upon common storage.
Journal of Industrial Engineering Chemistry 13 (-) 523-524;
1921
4

717 Gottwald, B.A.
Adsorption und zweidimensionale Kondensation an heterogenen
Oberflächen.
Zeitschrift für Physikalische Chemie 62 (-) 213-217; 1968
2

718 Gough, M.C.
The measurement of relative humidity, with particular
reference to remote long term measurement in grain silos.
Tropical Stored Products Information (27) 19-30; 1974
3

719 Gough, M.C.
Cinchona bark moisture relations.
Tropical Stored Products Information (28) 4-5; 1974
3,5

720 Gough, M.C.
A simple technique for the determination of humidity
equilibria in particulate foods.
Journal of Stored Products Research 11 (3/4) 161-166; 1975
3,5

721 Gough, M.C.
Remote measurement of moisture content in bulk grain using an
air extraction technique.
Journal of Agricultural Engineering Research 21 (2) 217-219;
1976
3

722 Gough, M.C.; Bateman, G.A.
Moisture humidity equilibria of tropical stored produce.
Part I Cereals.
Tropical Stored Products Information (33) 25-40; 1977
5

723 Gough, M.C.; King, P.E.
Moisture content relative humidity equilibria of some tropical
cereal grains.
Tropical Stored Products Information (39) 13-17; 1980
5

724 Gough, M.C.; Lippiat, G.A.
Moisture humidity equilibria of tropical stored produce.
Part II Oilseeds.
Tropical Stored Products Information (34) 49-61; 1977
5

725 Gough, M.C.; Lippiat, G.A.
Moisture humidity equilibria of tropical stored produce.
Part III Legumes, Spices and Beverages.
Tropical Stored Products Information (35) 15-29; 1978
5

726 Gough, M.C.; Wright, S.P.D.
Selected bibliography on the movement of water and heat in
cereals.
Tropical Stored Products Information (33) 17-20; 1977
2

727 Grau, R.; Hamm, R.
Über das Wasserbindungsvermögen des Säugetiermuskels.
II. Mitteilung: Über die Bestimmung der Wasserbindung des
Muskels.
Zeitschrift für Lebensmittel-Untersuchung und -Forschung
105 (-) 446-460; 1957
2

728 Greaves, P.
Water activity in foodstuffs.
Food Trade Review 46 (5) 281-282; 1976
1

729 Grecz, N.; Smith, R.F.; Hoffmann, C.C.
Sorption of water by spores, heat killed spores and vegetative
cells.
Canadian Journal of Microbiology 16 (-) 573-579; 1970
4
730 Greenberg, D.M.
Certain chemical and physical characteristics of the proteins.
In: Schmidt, C.L.A. (Ed.).
Chemistry of the amino acids and proteins.
Charles C. Thomas, Springfield, Baltimore; 1938
5
731 Greenewalt, C.H.
Partial pressure of water out of aqueous solutions of sulfuric
acid.
Industrial and Engineering Chemistry 17 (-) 522-523; 1925
3
732 Greenspan, L.
Low frost-point humidity generator.
Journal of Research of the National Bureau of Standards
77A (5) 671-677; 1973
3
733 Greenspan, L.
Humidity fixed points of binary saturated aqueous solutions.
Journal of Research of the National Bureau of Standards
Physics and Chemistry 81A (1) 89-96; 1977
3
734 Gregg, S.J.
An electrical sorption balance.
Journal of the Chemical Society (-) 561-562; 1946
3
735 Gregg, S.J.; Sing, K.
Adsorption, surface area and porosity.
Academic Press, London; 1967
3,4
736 Gregg, S.J.; Wintle, M.F.
An automatically recording electrical sorption balance.
Journal of Scientific Instruments 23 (-) 259-264; 1946
3
737 Greig, R.I.W.
Loss of water holding capacity of heat denatured whey protein.
Milk Industry 81 (3) 18-19, 21; 1979
2
738 Greig, R.I.W.
Effect of dehydration on water binding in fresh denatured whey
protein curd.
Dairy Industries International - (2) 5-14; 1979
2
739 Greig, R.I.W.
Sorption properties of heat denatured cheese whey protein.
Part 1 Moisture sorption isotherms.
Dairy Industries International 44 (5) 18-20,22, 24; 1979
Part 2 Unfreezable water content.
Dairy Industries International 44 (6) 15-17; 1979
Part 3 Porosity and pore size distribution
Dairy Industries International 44 (10) 17-19, 21; 1979
2,5
740 Greyson, J.; Levi, A.A.
Calorimetric measurements of the heat of sorption of water
vapor on dry swollen cellulose.
Journal of Polymer Science part A, 1 (-) 3333-3342; 1963
2,5

741 Grierson, W.; Wardowski, W.F.
Relative humidity effects on the postharvest life of fruits
and vegetables.
Hortscience 13 (5) 570-574; 1978
4

742 Griffin, W.C.
Hygroscopicity of softened glue composition.
Industrial and Engineering Chemistry 37 (11) 1126-1130; 1945
5

743 Grinberg, N.Kh.; Popovskii, V.G.; Krenis, G.A.;
Kerdivarenko, M.A.
Research into the sorption properties of soft fruit during
freeze-drying.(Orig.russ.)
Izvestiya Vysshikh Uchebnykh Zavedenii, Pishchevaya
Tekhnologiya (5) 113-115; 1972
2,5

744 Grinberg, N.Kh.; Popovskii, V.G.; Voloshina, M.Kh.
Determination of moisture content in freeze-dried products by
sorption isotherms.(Orig.russ.)
Konservnaya i Ovoshchesushil'naya Promyshlennost' 26 (3) 33-34;
1971
5

745 Gröninger, K.G.
Die Verwendung und Aufnahme von Sorptionsisothermen.
Mitteilung aus dem Gebiete der Lebensmitteluntersuchung und
-Hygiene 56 (4) 255-264; 1965
1,3

746 Gröninger, K.G.
The equilibrium moisture measurement of foodstuffs.
In: Relative Humidity in the Food Industry.
The British Food Manufacturing Industries Research Association
Symposium Proc. No. 4, pp 16-27. Leatherhead, London; 1969
3,5

747 Gröninger, K.G.
Ein neues Verfahren zur kontinuierlichen
Feuchtigkeits-Überwachung getrockneter Nahrungsmittel.
Mitteilungen aus dem Gebiete der Lebensmitteluntersuchung und
-Hygiene 59 (-) 460-470; 1969
3

748 Gröninger, K.G.
The equilibrium moisture measurement of foodstuffs.
Food Trade Review 41 (4) 31-32; 1971
3

749 Gronau, J.
Untersuchungen zur Deutung von Wasserdampfsorptionsisothermen
unter Berücksichtigung der Schwingungserscheinungen.
Silikattechnik 20 (1) 21-24; 1969
1,2

750 Grosch, W.
Wasseraktivität und Oxydation von Lipiden.
Lebensmitteltechnologie 12 (4) 11-18; 1979
4

751 Grover, D.W.
The keeping properties of confectionery as influenced by its
water vapour pressure.
Journal of the Society of Chemical Industry 66 (-) 201-205;
1947
4,5

752 Grover, D.W.; Nicol, J.M.
The vapor pressure of glycerin solutions at 20 °C.
Journal of the Society of Chemical Industry; 59 (-) 175-177;
1940
3

753 Guardia, E.J.; Haas, G.J.
Influence of water binders on the activity and thermal
inactivation of lipase.
Journal of Agricultural and Food Chemistry 15 (-) 412; 1967
4,5

754 Gudkov, A.V.; Fedin, F.A.
Dependence of water activity in cheeses on salt and moisture
content.(Orig.russ.)
Trudy, Vsesoyuznyi Nauchno-Issledovatel'skii Institut
Maslodel'noi i Syrodel'noi Promyshlennosti 11 (-) 30-33; 1973
3,5

755 Gudkov, A.V.; Fedin, F.A.
Effect of water activity on development of lactic acid
bacteria.(Orig.russ.)
Molochnaya Promyshlennost' 10 (-) 20-23, 46; 1978
4

756 Guerzoni, M.E.; Suzzi, G.; Lerici, C.R.; Bartolini, R.;
Testa, G.
Water activity and food stability. I. Effects on viability of
Saccharomyces cerevisiae cells.(Orig.ital.)
Rivista di Scienza e Tecnologia degli Alimenti e di Nutrizione
Umana 6 (5/6) 295-300; 1976
4

757 Guex, M.M.
Some combined reactions of water at the surface of freeze
dried products.
In: Goldblith, S.A.; Rey, L.; Rothmayr, W.W. (Edts.).
Freeze Drying and Advanced Technology. pp 413-420.
Academic Press, London; 1975
4,5

758 Guggenheim, E.A.
Applications of statistical mechanics.
Clarendon Press, Oxford; 1966
1,2

759 Guggenheim, E.A.
Thermodynamics. 5. ed.
North-Holland Publ. Co., Amsterdam; 1967
2

760 Guilbert, S.; Clément, O.; Cheftel, J.C.
Efficacité comparée d'agents dépresseurs de l'aw en solution
et dans des aliments a humidité intermédiaire.
Lebensmittel-Wissenschaft und -Technologie 14 (5) 245-251; 1981
4,5

761 Guilbot, A.
Le problème de l'eau "liée" dans les colloides biologiques.
Industries Alimentaires et Agricoles 1 (1) 7-13; 1952
2

762 Guilbot, A.
Activités enzymatiques en milieu peu hydraté.
Annales de la Nutrition et de l'Alimentation 21 (4) B87-B112;
1967
4

763 Guilbot, A.; Charbonniere, R.; Abadie, P.; Girard, P.
L'eau de sorption de l'amidon: Etude par spectrologie
hertzienne.
Die Stärke 12 (11) 327-332; 1960
2,5

764 Guilbot, A.; Drapron, R.
Evolution, en fonction de l'humidité relative, de l'état
d'organisation et de l'affinité pour l'eau, de divers
oligosides cryodeshydratés.
Annexe 1969-9 au Bulletin de l' Institut International de Froid.
pp 191-195. Paris; 1969
4

765 Guilbot, A.; Guillemet, R.
L'eau liée dans les composants de la pâte de farine de blé.
Etude de quelques facteurs suscpetibles de l'influencer.
Bulletin de la Societé de Chimie Biologique 27 (7-9) 335-341;
1945
2

766 Guilbot, A.; Lindenberg, A.B.
Eau non solvante et eau de sorption de la cellule de levure.
Biochemica et Biophysica Acta 39 (-) 389-397; 1960
2,5

767 Gulstad, C.; Heitke, T.; Dreier, W.
Shelf stable, intermediate moisture doughs.
U.S. Patent Nr. 3767421; 1973
4

768 Gunn, R.D.; King, C.J.
Mass transport characteristics of freeze-dried foods.
Chemical Engineering Progress Symposium Series
67 (108) 94-101; 1971
2

769 Gupta, S.L.; Bhatia, R.K.S.
Sorption of water and organic vapors on starch at 35°C.
Indian Journal of Chemistry 7 (-) 1231-1233; 1969
2,5

770 Gupta, S.L.; Bhatia, R.K.S.
Influence of certain modifications of starch on its sorptive
properties with water vapor.
Indian Journal of Chemstry 8 (-) 536-540; 1970
2,5

771 Gur-Arieh, C.
Moisture adsorption of wheat flour as affected by physical and
chemical characteristics.
Ph.D. Thesis, University of Illinois; 1963
1,2,3,4,5

772 Gur-Arieh, C.; Nelson, A.I.; Steinberg, M.P.; Wei, L.S.
A method of rapid determination of moisture adsorption
isotherms of solid particles.
Journal of Food Science 30 (1) 105-110; 1965
3,5

773 Gur-Arieh, C.; Nelson, A.I.; Steinberg, M.P.; Wei, L.S.
Water activity of flour at high moisture contents as measured
wih a pressure membrane cell.
Journal of Food Science 30 (2) 188-191; 1965
2,3,5

774 Gur-Arieh, C.; Nelson, A.I.; Steinberg, M.P.
Studies on the density of water adsorbed on low-protein
fraction of flour.
Journal of Food Science 32 (-) 442-445; 1967
2

775 Gur-Arieh, C.; Nelson, A.I.; Steinberg, M.P.; Wei, L.S.
Moisture adsorption by wheat flours and their cake baking
performance.
Food Technology 21 (3) 412-415; 1967
5

776 Gustafson,R.J.
     Equlibrium moisture content of shelled corn from 50 to 155°F.
     Thesis, Urbana, University of Jllinois; 1972
     2,3,5

777 Gustafson, R.J.; Hall, G.E.
     Equilibrium moisture content of shelled corn from 50 to 155°F
     Transactions of the American Society of Agricultural Engineers
     16 (-) 120-124; 1974
     2,3,5

778 Haas, G.J.; Bennett, D.; Herman, E.B.; Colette, D.
     Microbial stability of intermediate moisture foods.
     Food Product Development, Chicago, 9 (3) 86-94; 1975
     4

779 Haas, G.J.; Herman, E.B.
     Bacterial growth in intermediate moisture food systems.
     Lebensmittel-Wissenschaft und -Technologie 11 (2) 74-78; 1978
     4

780 Hadorn, H.
     Ringversuche zur Wasserbestimmung in Lebensmitteln nach Karl
     Fischer.
     Mitteilungen aus dem Gebiete der Lebensmitteluntersuchung und
     Hygiene 71 (-) 221-235; 1980
     3

781 Hadorn, H.; Keme, T.; Kleinert, J.; Messerli, M.; Zürcher, K.
     So verhalten sich Haselnüsse unter verschiedenen
     Lagerungsbedingungen.
     Zucker- und Süßwaren-Wirtschaft 30 (4) 120-126, (5) 170-180;
     1977
     4,5

782 Hadorn, H.; Keme, T.; Kleinert, J.; Messerli, M.; Zürcher, K.
     Lagerungsversuche und Qualitätsprüfungen an Haselnüssen.
     Gordian 78 (10) 300-310; 1978
     4

783 Hadorn, H.; Zürcher, K.
     Nachteilige Veränderungen von Haselnüssen während der
     Lagerung.
     Gordian 77 (5) 114-120; 1977
     4

784 Haegerdal, B.; Loefqvist, B.
     Screening method based on electric hygrometer for obtaining
     water sorption isotherms.
     Journal of Agricultural and Food Chemistry 21 (3) 445-451;
     1973
     3,5

785 Haegerdal, B.; Martens, H.
     Influence of water content on the stability of myoglobin to
     heat treatment.
     Journal of Food Science 41 (-) 933-937; 1976
     4

786 Härnulv, B.; Johansson, M.; Snygg, B.
     Heat resistance of Bacillus stearothermophilus spores at
     different water activities.
     Journal of Food Science 42 (1) 91-93; 1977
     4

787 Härnulv, B.; Snygg, B.G.
     Heat resistance of Bacillus subtilis spores at various water
     activities.
     Journal of Applied Bacteriology 35 (4) 615-624; 1972
     4

788 Hagenmaier, R.
     Water binding of some purified oilseed proteins.
     Journal of Food Science 37 (-) 965-966; 1972
     2

789 Hagenmaier, R.D.; Cater, C.M.; Mattil, K.F.
Coconut skim milk as an intermediate moisture product.
Journal of Food Science 40 (4) 717-720; 1975
4,5

790 Hagymassy, J.; Brunauer, S.; Mikhail, R.
Pore structure analysis by water vapor adsorption.
1. Curves for water vapor.
Journal of Colloid and Interface Science 29 (3) 485-491; 1969
2

791 Hailwood, A.J.; Horrobin, S.
Absorption of water by polymers: Analysis in terms of a simple
model.
Transactions of the Faraday Society 42 B, 84-92; 1946
2

792 Hall, C.W.
Problems of humidity and moisture in agriculture.
In: Wexler, A.; Amdur, E.J. (Edts.).
Humidity and Moisture. Measurement and Control in Science and
Industry. Vol. 2, pp 87-94. Reinhold Publ. Corp., New York; 196!
4,5

793 Hall, C.W.
Equilibrium moisture content of cocoa beans.
Acta Agronomica Palmira 10 (1) 53-56; 1960
cited in: Gough,M.C.;Lippiatt,G.A.
Tropical Stored Products Information (35) 15-29; 1978
5

794 Hall, C.W.; Rodriguez-Arias, J.H.
Equilibrium moisture content of shelled corn.
Agricultural Engineering 39 (8) 466-470; 1958
1,2,5

795 Halsey, G.D.
Physical adsorption on non-uniform surfaces.
Journal of Chemical Physics 16 (10) 931-937; 1948
2

796 Hamano, M.; Aoyama, Y.; Yokotsuka, T.
Moisture sorption of powdered soy sauce.(Orig.jap.)
Journal of the Japan Society of Food Science and Technology
(Nihon shokuhin Kogyo Gakkai-shi) 19 (11) 503-507; 1972
2,5

797 Hamano, M.; Aoyama, Y.
Effect of saturated fatty acids on hygroscopic equilibria of
spray dried soy sauce.(Orig.jap.)
Journal of the Agricultural Chemical Society of Japan (Nihon
Nogei Kagakkai-shi) 47 (11) 719-725; 1973
5

798 Hamano, M.; Aoyama, Y.
Changes of components and behavior of water sorption during
spray drying of soy sauce.(Orig.jap.)
Journal of the Agricultural Chemical Society of Japan 4(Nihon
Nogei Kagakkai-shi) 48 (11) 619-625; 1974
2

799 Hamano, M.; Aoyama, Y.; Sugimoto, H.
Effect of sugar on water sorption of powdered soy sauce.
(Orig. jap.)
Journal of the Agricultural Chemical Society of Japan (Nihon
Nogei Kagakkai-shi) 50 (7) 401-407; 1976
5

800 Hamano, M.; Sugimoto, H.
Water sorption, reduction of caking and improvement of free
flowingness of powdered soy sauce and miso.
Journal of Food Processing and Preservation 2 (3) 185-196;
1978
5

801 Hamm, R.
     Biochemistry of meat hydration.
     Advances in Food Research 10 (-) 355-463; 1960
     4,5
802 Hamm, R.
     Beziehungen zwischen dem Wasserbindungsvermögen des Fleisches
     und sensorischen Faktoren der Fleischqualität.
     Lebensmitteltechnologie 11 (4) 8-19; 1978
     2,4
803 Hamm, R.; Potthast, K.; Acker, L.
     Einfluß der Wasseraktivität auf enzymatische Veränderungen in
     gefriergetrocknetem Muskelfleisch.
     Zeitschrift für Lebensmittel-Untersuchung und -Forschung
     154 (2) 73-79; 1974
     4
804 Han, S.B.; Lee, J.H.; Lee, K.H.
     Non-enzymatic browning reactions in dried anchovy when stored
     at different water activities.(Orig.kor.)
     Bulletin of the Korean Fisheries Society 6 (1/2) 37-43; 1973
     4
805 Hanoussek, J.
     Guarantee terms for packaged hygroscopic substances.(Orig.czech.)
     Prumysl Portravin 12 (-) 255-260; 1961
     2,4
806 Hansemann, J.Y.; Guilbert, S.; Richard, N.; Cheftel, J.C.
     Influence conjointe de l'activité de l'eau et du pH sur la
     croissance de Staphylococcus aureus dans un aliment carne à
     humidité intermédiaire.
     Lebensmittel-Wissenschaft und -Technologie 13 (5) 269-270;
     1980
     4
807 Hansen, J.R.
     Hydration of soybean protein.
     Journal of Agricultural and Food Chemistry 24 (-) 1136-1141;
     1976
     2,5
808 Hansen, J.R.
     Hydration of soybean protein. 2.Effect of isolation method and
     various other parameters on hydration.
     Journal of Agricultural and Food Chemistry 26 (2) 301;1978
     2,5
809 Hansen, G.P.; Gough, M.C.
     Moisture characteristics of Macadamia nuts in relation to
     storage.
     Journal of the Science of Food and Agriculture 28 (11) 990-995;
     1977
     4,5
810 Hansen, N.H.; Riemann, H.
     Mikrobiologische Beschaffenheit von vorverpacktem Fleisch und
     vorverpackten Fleischwaren.
     Die Fleischwirtschaft 14 (9) 861-868; 1962
     4
811 Hanson, S.W.F. (Ed.).
     The accelerated freeze drying (AFD) method of food
     preservation.
     Her Majesty's Stationery Office (H.M.S.O.), London; 1961
     2,4,5
812 Hanson, T.P.; Cramer, W.D.; Abraham, W.H.; Lancaster, E.B.
     Rates of water-vapour absorption in granular cornstarch.
     Chemical Engineering Progress Symposium Series 67 (108) 35-39;
     1971
     3,4,5

813 Hardman, T.M.
Measurement of water activity. Critical appraisal of methods.
In: Davies, R.; Birch, G.G.; Parker, K.J. (Edts.).
Intermediate Moisture Foods. pp 75-88.
Applied Science Publ., London; 1976
3,5

814 Hardman, T.M.
The significant role played by water present in foodstuffs.
In: King, R.D. (Ed.).
Developments in Food Analysis Techniques. Vol. 1, pp 75-124.
Applied Science Publ., London; 1978
2,3,4,5

815 Hardy, J.
L'activité de l'eau, le sel et les moisissures des fromages.
Revue Laitière Francaise 19 (377) 19-25; 1979
1,4,5

816 Harel, S.; Kranner, J.; Juven, B.J.; Golan, R.
Long-term preservation of high-moisture dried apricots with
and without chemical preservatives.
Lebensmittel-Wissenschaft und -Technologie 11 (4) 219-221;
1978
4,5

817 Harkins, W.D.; Jura, G.J.
An absolute method for the determination of the area of a fine
crystalline powder.
Journal of Chemical Physics 11 (9) 430; 1943
3

818 Harkins, W.D.; Jura, G.
A vapor adsorption method for the determination of the area of
a solid.
Journal of the American Chemical Society 66 (-) 1366-1373;
1944
3

819 Harris, C.
Determination of water.
Talanta 19 (-) 1523-1547; 1972
3

820 Hart, J.R.
Hysteresis effects in mixtures of wheats taken from the same
sample but having different moisture contents.
Cereal Chemistry 41 (9) 340-350; 1964
2,5

821 Harwig, J.; Blanchfield, B.J.; Jarvis, G.
Effect of water activity on disappearance of patulin and
citrinin from grains.
Journal of Food Science 42 (5) 1225-1228; 1977
4

822 Hasegawa, Y.; Sekine, T.
Studies on liquid - liquid partitions systems. The
determination of water activity by solvent extraction
technique.
Bulletin of the Chemical Society of Japan 38 (-) 1713-1716;
1965
3

823 Hauser, H.
Water-phospholipid interactions.
In: Duckworth, R.B. (Ed.).
Water Relations of Foods. pp 37-71.
Academic Press, London; 1975
4

824 Hauser, P.M.; McLaren, A.D.
Permeation through and sorption of water vapor by high
polymers.
Industrial and Engineering Chemistry 40 (1) 112-117; 1948
2,5

825 Hayakawa, K.I.
Predicting an equilibrium state value from transient state
data.
Journal of Food Science 39 (2) 272-275; 1974
2

826 Hayakawa, K.I.; Matas, J.; Hwang, M.P.
Moisture sorption isotherms of coffee products.
Journal of Food Science 43 (3) 1026-1027; 1978
2,5

827 Haynes, B.C. jr.
Vapor pressure determination of seed hygroscopicity.
Technical Bulletin No. 1229
Gov. Printing Office, Washington, D.C.
3,5

828 Heaton, E.K.; Beuchat, L.R.
Quality characteristics of high moisture pecans stored at
refrigeration temperatures.
Journal of Food Science 45 (1) 255-258, 261; 1980
4

829 Hedlin, C.P.
A device for calibration electrical humidity sensors.
Materials Research and Standards 6 (1) 25-29; 1966
3

830 Hedlin, C.P; Trofimenkoff, F.N.
Relative humidities over saturated solutions of nine salts in
the temperature range from 0 to 90 °F.
In: Wexler, A.; Wildhack, W.A. (Edts.). Humidity and Moisture.
Vol. 3, pp 519-520. Reinhold Publ. Corporation, New York; 1965
3

831 Heidelbaugh, N.D.; Goldblith, S.A.
Some stability and safety aspects of intermediate moisture
foods.
In: Goldblith, S.A.; Rey, L.; Rothmayr, W.W. (Edts.).
Freeze Drying and Advanced Technology. pp 675-689.
Academic Press, London; 1975
4

832 Heidelbaugh, N.D.; Karel, M.
Effect of water binding agents on the catalyzed oxidation of
methyllinoleate.
Journal of the American Oil Chemists Society 47 (-) 539-544;
1970
4

833 Heidelbaugh, N.D.; Karel, M.
Intermediate moisture food technology.
In: Goldblith, S.A.; Rey, L.; Rothmayr, W.W. (Edts.).
Freeze Drying and Advanced Food Technology. pp 619-641.
Academic Press, London; 1975
4

834 Heidelbaugh, N.D.; Yeh, C.P.; Karel, M.
Effects of model system composition on autoxidation of methyl
linoleate.
Journal of Agricultural and Food Chemistry 19 (-) 140-142;
1971
4

835 Heinevetter, L.; Kroll, J.; Gassmann, B.; Hoppe, K.
    Zur Bestimmung der Wasseraufnahmefähigkeit pulverförmiger
    Proteinrohstoffe.
    1. Mitt. Die Kapillarsaugmethode.
    Die Nahrung 27 (6) 557-567; 1983
    2,3

836 Heinrich, C.
    Thermogravimetrische Untersuchungen zur Wassergehaltsbestimmung
    von Milchpulver.
    Milchwissenschaft 25 (7) 387-391; 1970
    3

837 Heintzeler, I.
    Das Wachstum der Schimmelpilze in Abhängigkeit von den
    Hydrataturverhältnissen unter verschiedenen Außenbedingungen.
    Archiv für Mikrobiologie 10 (-) 92-132; 1939
    4

838 Heiss, R.
    Die Verpackung feuchtigkeitsempfindlicher Lebensmittel.
    I. Mitteilung: Grundlagen für die Entwicklung wirtschaftlicher
    Verpackungen.
    Zeitschrift für Lebensmittel-Untersuchung und -Forschung
    89 (-) 173-183; 1949
    4,5

839 Heiss, R.
    Untersuchungen über die Haltbarkeit von Hartkaramellen.
    II. Mitteilung: Die Berechnung der Sorptionsisothermen für
    Hartkaramellen.
    Die Stärke 7 (3) 45-51; 1955
    5

840 Heiss, R.
    Untersuchungen über die Haltbarkeit von Hartkaramellen.
    III. Mitteilung: Zeitabhängigkeit des Dampfdruckes über der
    Bonbonoberfläche.
    Die Stärke 7 (3) 51-55; 1955
    2,3

841 Heiss, R.
    Haltbarkeit von Backpulver. I. Mitteilung.
    Die Stärke 7 (8) 209-216; 1955
    4,5

842 Heiss, R.
    Untersuchungen über die Haltbarkeit verpackter
    feuchtigkeitsempfindlicher Güter.
    Chemie-Ingenieur-Technik 28 (12) 763-768; 1956
    4,5

843 Heiss, R.
    Die Verpackung feuchtigkeitsempfindlicher Güter.
    Springer-Verlag, Berlin, Göttingen, Heidelberg; 1956
    1,2,4

844 Heiss, R.
    Haltbarkeit und Sorptionsverhalten wasserarmer Lebensmittel.
    Springer-Verlag, Berlin, Heidelberg, New York; 1968
    1,4,5

845 Heiss, R.
    Einfluß der Verpackung auf die Qualität von Lebensmitteln
    (aufgezeigt an wasser- und sauerstoffempfindlichen
    Lebensmitteln).
    In: Proc. of the 6th European Symposium Food Engineering
    and Foodquality, Cambridge, 8. - 10.09.1975; pp 281-295.
    Society of Chemical Industry, London; 1975
    1,4

846   Heiss, R.
      Grundlagen der Verpackung wasserdampf-, sauerstoff- und
      lichtempfindlicher Lebensmittel.
      Lebensmittel-Wissenschaft und -Technologie 10 (3) 123-130; 1977
      1,4,5
847   Heiss, R.
      Verpackung von Lebensmitteln.
      Springer Verlag, Berlin, Heidelberg, New York; 1980
      1,4
848   Heiss, R.; Eichner, K.
      Moisture content and shelflife.
      Food Manufacture 46 (5) 53-65, (6) 37-42; 1971
      1,4
849   Heiss, R; Eichner, K.
      Die Haltbarkeit von Lebensmitteln mit niedrigen und mittleren
      Wassergehalten.
      Chemie Mikrobiologie Technologie der Lebensmittel 1 (1) 33-40;
      1971
      1,4
850   Heiss, R.; Schachinger, L.
      Untersuchungen über die Haltbarkeit von Hartkaramellen.
      I. Vorgänge in der Randschicht.
      Die Stärke 5 (6) 152-157; 1953
      4,5
851   Heitz, F.
      Théorie de l'activité du saccharose en solution aqueuse.
      La Sucrerie Belge 90 (8) 383-396; 1971
      2,5
852   Heldman, D.R.; Hall, C.W.; Hedrick, T.I.
      Vapor equilibrium relationships of dry milk.
      Journal of Dairy Science 48 (7) 845-852; 1965
      2,5
853   Heldman, D.R.; Hall, C.W.; Hedrick, T.I.
      Equilibrium moisture contents and moisture adsorption rates of
      dry milks.
      In: Wexler, A.; Amdur, E.J. (Edts.).
      Humidity and Moisture. Measurement and Control in Science and
      Industry. Vol. 2, pp 173-184. Reinhold Publ. Corporation,
      New York; 1965
      2,5
854   Heldman, D.R.; Reidy, G.; Palnitkar, M.
      Texture stability during storage of freeze dried beef at low
      and intermediate moisture contents.
      Journal of Food Science 38 (2) 282-285; 1973
      4
855   Hellman, N.N.; Boesch, T.F.; Melvin, E.H.
      Starch granule swelling in water vapor sorption.
      Journal of the American Chemical Society 74 (1) 348-350; 1952
      4,5
856   Hellman, N.N.; Melvin, E.H.
      Water sorption by corn starch as influenced by preparatory
      procedures and storage time.
      Cereal Chemistry 25 (-) 146-150; 1946
      4,5
857   Hellman, N.N.; Melvin, E.H.
      Surface area of starch and its role in water sorption.
      Journal of the American Chemical Society 72 (11) 5186-5188;
      1950
      2,3,5

858 Hendel, C.E.; Burr, H.K.; Boggs, M.H.
Factors affecting storage stability of potato granules.
U.S. Bureau of Agriculture and Industrial Chemistry.
AIC-303; 1951
4
859 Hendel, C.E.; Silveira, V.G.; Harrington, W.O.
Rates of non enzymatic browning of white potato during
dehydration.
Food Technology 9 (-) 433-438; 1955
2,4
860 Henderson, S.M.
A basic concept of equilibrium moisture.
Agricultural Engineering 33 (1) 29-32; 1952
2,5
861 Henderson, S.M.
Equilibrium moisture content of small grain hysteresis.
Transactions. American Society of Agricultural Engineers
13 (6) 762-764; 1970
2,3,5
862 Henderson, S.M.
Equilibrium moisture content of hops.
Journal of Agricultural Engineering Research 18 (1) 55-58; 1973
5
863 Henderson, S.M.; Pixton, S.W.
The adsorption of moisture by spray-dried skimmed milk.
Journal of Stored Products Research 16 (-) 47-49; 1980
5
864 Henderson, S.M.; Pixton, S.W.
The influence of the testa on sorption of water by cocoa beans
and some legumes.
Journal of Stored Products Research 16 (2) 81-84; 1980
5
865 Henderson, S.M.; Pixton, S.W.
Relative humidity control in equilibration studies.
Journal of Science of Food and Agriculture 32 (-) 1145-1150;
1981
2,3
866 Hendrickx, H.; Moor, H. de
Moisture absorption by dried milk and dried whey during
storage.
Mededelingen van de Rijksfaculteit Landbouwwetenschappen te
Gent 34 (4) 1089-1103; 1969
2,4
867 Hennig, H.J.; Lechert, H.; Krische, B.
Kernresonanz-Untersuchungen zum Einfluß der technischen
Aufarbeitung auf die Wasserbindungsfähigkeit von
Kartoffelstärke.
Die Stärke 27 (5) 151-154; 1975
2
868 Hennig, H.J.; Lechert, H.
Temperaturabhängigkeit des kernmagnetischen
Relaxationsverhaltens von Wasserprotonen in nativer Stärke.
Zeitschrift für Naturforschung 31a (-) 306-309; 1976
2
869 Hennig, H.J.; Lechert, H.; Goemann, W.
Untersuchung des Quellverhaltens von Stärke mit Hilfe der
Kernresonanz-Impuls-Spektroskopie.
Die Stärke 28 (1) 10-13; 1976
2,4

870 Henry, K.M.; Kon, S.K.; Lea, C.H.; White, J.C.D.
Deterioration on storage of dried skim milk.
Journal of Dairy Research 15 (-) 292-363; 1948
4

871 Hepburn, J.R.
The vapour pressure of water over sulphuric acid-water
mixtures and its measurement by an improved dew point
apparatus.
Proc. of the Physical Society, London 40 (-) 249-260; 1928
3

872 Hermansson, A.M.
Functional properties of proteins for foods-swelling.
Lebensmittel-Wissenschaft und -Technologie 5 (1) 24-29; 1972
4

873 Hermansson, A.M.
Functional properties of added proteins correlated with
properties of meat systems. Effect of texture of a meat
product.
Journal of Food Science 40 (-) 611-614; 1975
4

874 Hermansson, A.M.
Functional properties of proteins for foods - water vapour
sorption.
Journal of Food Technology 12 (2) 177-187; 1977
1,5

875 Herrmann, J.
Lehrbuch der Vorratspflege.
Deutscher Landwirtschaftsverlag, Berlin; 1963
4,5

876 Herrmann, J.
Die Haltbarkeit von Lebensmitteln als Wechselbeziehung
zwischen Ausgangsqualität, Verpackung, Transport und
Lagerbedingungen.
Die Nahrung 18 (4) 409-424; 1974
4,5

877 Herrmann, J.; Vogel, J.; Köller, M.
Die Aktivität von Glucoseoxydase in Abhängigkeit von
Temperatur, Wassergehalt und pH-Wert.
Die Nahrung 9 (6) 659-667; 1965
4

878 Herrmann, W.
Die Adsorption von Wasserdampf in Schwerspatpresslingen und
ihr Einfluß auf deren Festigkeit.
Dissertation, Universität Karlsruhe; 1971
3

879 Herrmann, W.
Die Beeinflussung der Festigkeit von Preßlingen durch die
Adsorption von Wasserdampf.
Chemie-Ingenieur-Technik 43 (7) 418-425; 1971
3,4

880 Hertl, W.; Hair, M.L.
Adsorption of water on silica.
Nature (London) 223 (5211) 1150-1151; 1969
5

881 Heskestad, R.; Steinsholt, K.
Water activity in foods.
Meieriposten 67 (11) 327-329,331-332; 1978
1,5

882 Hickman, M.J.
     Measurement of humidity.
     National Physical Laboratory Notes on Applied Science
     4 (-) 31; 1970
     3,5
883 Hill, J.E.; Sunderland, J.E.
     Equilibrium vapor pressure and latent heat of sublimation for
     frozen meats.
     Food Technology 21 (9) 1276-1278; 1967
     2,5
884 Hill, P.E.; Rizvi, S.S.H.
     Thermodynamic parameters and storage stability of drum dried
     peanut flakes.
     Lebensmittel-Wissenschaft und -Technologie 15 (4) 185-190; 1982
     2,4,5
885 Hill, T.L.
     Statistical mechanics of adsorption.
     V. Thermodynamics and heat of adsorption.
     IX. Adsorption thermodynamics and solution thermodynamics.
     Journal of Physical Chemistry 17 (-) 520-535; 1949
     Journal of Physical Chemistry 18 (3) 246-256; 1950
     2
886 Hill, T.L.
     Introduction to statistical thermodynamics.
     Addison-Wesley Publ. Co., Reading, Mass.; 1960
     1,2
887 Hill, T.L.; Rowen, J.W.
     Sorption of vapors by polymers.
     Journal of Polymer Science 9 (1) 93-95; 1952
     2
888 Hirschberg, F.
     Investigations into the osmotic resistance of plant cell walls
     and skins by means of equilibrium relative humidity
     determinations.
     Acta Alimentaria Academiae Scientarium Hungaricae 3 (1) 27-35;
     1974
     2,3
889 Hirschberg, F.; Szabo, Z.
     Measurement of water activity of biological substances and
     introduction of the concept of steric "hydrature surface".
     Acta Alimentaria Academiae Scientarium Hungaricae
     1 (3/4) 355-369; 1972
     2,3,5
890 Hnojewyj, W.S.; Reyerson, L.H.
     The sorption of H2O and D2O by lyophilized lysozyme.
     Journal of Physical Chemistry 63 (-) 1653-1654; 1959
     2,5
891 Hnojewyj, W.S.; Reyerson, L.H.
     Further studies on the sorption of H2O and D2O vapors by
     lysozyme and the deuterium-hydrogen exchange effect.
     Journal of Physical Chemistry 65 (10) 1694-1698; 1961
     2,5
892 Hofer, A.A.
     Zur Aufnahmetechnik von Sorptionsisothermen und ihre Anwendung
     in der Lebensmittelindustrie.
     Inauguraldissertation, Universität Basel; 1962
     1,5
893 Hofer, A.A.; Mohler, H.
     Apparatur zur Messung der Sorptionskinetik und eine
     Mikromethode zur Messung von Sorptionsisothermen.
     Helvetica Chimica Acta 45 (5) 1415-1418; 1962
     3,5

894 Hofer, A.A.; Mohler, H.
Zur Aufnahmetechnik von Sorptionsisothermen und ihre Anwendung
in der Lebensmittelindustrie.
Mitteilungen aus dem Gebiete der Lebensmitteluntersuchung und
Hygiene 53 (-) 274-290; 1962
1,3,5

895 Hofmann, K.
Ein neues Gerät zur Bestimmung der Wasserbindung des
Fleisches: Das "Kapillar-Volumeter".
Die Fleischwirtschaft 27 (1) 25-30; 1975
2

896 Hofmann, K.
Die Wasserbindung des Fleisches und ihre Messung.
Die Fleischwirtschaft 29 (4) 727-731; 1977
2

897 Hogan, J.T.; Karon, M.L.
Hygroscopic equilibria of rough rice at elevated temperatures.
Journal of Agricultural and Food Chemistry 3 (10) 855-860;
1955
2,3,5

898 Hollenbeck, R.G.; Peck, G.E.; Kildsig, D.O.
Application of immersional calorimetry to investigation of
solid-liquid interactions: Microcrystalline cellulose - water
system.
Journal of Pharmaceutical Sciences 67 (11) 1599-1606; 1978
2,5

899 Holley, W.
Halbfeuchte Lebensmittel.
In: Loncin, M. (Ed.).
Hochschulkurs ausgewählte Themen der
Lebensmittelverfahrenstechnik "Wasseraktivität".
Institut für Lebensmittelverfahrenstechnik
University Karlsruhe; 1980
4

900 Hollis, F.; Kaplow, M.; Halik, J.; Nordstrom, H.
Parameters for moisture content for stabilization of food
products (phase II).
U.S. Army, Natick Laboratories Technical Report 70-12-FL; 1969
4

901 Holm, F.
Physical characteristics of an air classified potato starch.
Journal of Food Technology 16 (2) 101-113; 1981
5

902 Holmquist, G.U.; Walker, H.W.; Stahr, H.M.
Influence of temperature, pH, water activity and antifungal
agents on growth of Aspergillus flavus and A. parasiticus.
Journal of Food Science 48 (3) 778-782; 1983
4

903 Hornung, E.W.; Giauque, W.F.
The vapor pressure of water over aqueous sulfuric acid at 25°C.
Journal of the American Chemical Society 77 (-) 2744-2746;
1955
3

904 Hoover, S.R.; Mellon, E.F.
Application of polarization theory to sorption of water vapor
by high polymers.
Journal of the American Chemical Society 72 (-) 2562-2566;
1950
2,5

905 Hopf, L.
Trocknung und Lagerung von Getreide in den europäischen
Ländern.
Die Müllerei 8 (-) 35; 1955
5

906 Horner, K.J.; Anagnostopoulos, G.D.
Combined effects of water activity, pH and temperature on the
growth and spoilage potential of fungi.
Journal of Applied Bacteriology 36 (-) 427-436; 1973
4

907 Horner, K.J.; Anagnostopoulos, G.D.
Effect of water activity on heat survival of Staphylococcus
aureus, Salmonella typhimurium and Salmonella senftenberg.
The Journal of Applied Bacteriology 38 (1) 9-17; 1975
4

908 Hottenroth, B.
Über die Hitzeinaktivierung der Lipase und anderen
Carboxylester-Hydrolasen in fettreichen Vollkonserven.
Die Fleischwirtschaft 26 (6) 1071-1075; 1974
5

909 Houston, D.F.
Hygroscopic equilibrium of brown rice.
Cereal Chemistry 29 (1) 71-76; 1952
3,5

910 Houston, D.F.; Kester, E.B.
Hygroscopic equilibria of whole-grain edible forms of rice.
Food Technology 8 (6) 302-304; 1954
3,5

911 Houston, D.F.; Straka, R.P.; Hunter, I.R.; Roberts, R.L.;
Kester E.B.
Changes in rough rice of different moisture content during
storage at controlled temperatures.
Cereal Chemistry 34 (6) 444-456; 1957
4

912 Hsieh, F.; Acott, K.; Elizondo, H.; Labuza, T.P.
The effect of water activity on the heat resistance of
vegetative cells in the intermediate moisture range.
Lebensmittel-Wissenschaft und -Technologie 8 (2) 78-81; 1975
4

913 Hsieh, F.; Acott, K.; Labuza, T.P.
Prediction of microbial death during drying of a macaroni
product.
Journal of Milk and Food Technology 39 (9) 619-623; 1976
4

914 Hsu, W.H.; Deng, J.C.
Processing of cured mullet roe.
Journal of Food Science 45 (1) 97-101; 1980
4

915 Hsu, W.H.; Deng, J.C.; Cornell, J.A.
Effect of salting time, dehydration temperatures and
dehydration time on quality of intermediate moisture mullet
roe.
Journal of Food Science 45 (1) 102-106; 1980
4

916 Hsu, H.W.; Deng, J.C.; Koburger, J.A.; Cornell, J.A.
Storage stability of intermediate moisture mullet roe.
Journal of Food Science 48 (1) 172-175, 196; 1983
4,5

917 Hubbard, J.E.; Earle, F.R.; Senti, F.R.
Moisture relations in wheat and corn.
Cereal Chemistry 34 (11) 422-433; 1957
3,5

918 Hückel, E.
Adsorption und Kapillarkondensation.
Akademische Verlags-Gesellschaft, Leipzig; 1928
1,2

919 Hüttig, G.F.
Zur Auswertung der Adsorptions-Isothermen.
Monatshefte für Chemie und verwandte Teile anderer
Wissenschaften 78 (-) 177-184; 1948
2

920 Huffman, V.L.; Lee, C.K.; Burns, E.E.
Selected functional properties of sunflower meal (Helianthus
annus).
Journal of Food Science 40 (-) 70-74; 1975
4,5

921 Hughes, F.J.; Vaala, J.L.; Koch, R.B.
Rapid measurement of moisture in flour by hygrometry.
In: Wexler, A.; Amdur, E.J. (Edts.).
Humidity and Moisture. Measurement and Control in Science and
Industry. Vol. 2, pp 133-136. Reinhold Publ. Corporation,
New York; 1965
2,3,5

922 Hukill, W.V.
Moisture in grain.
In: Wexler, A.; Amdur, E.J. (Edts.).
Humidity and Moisture. Measurement and Control in Science and
Industry. Vol. 2, pp 116-122. Reinhold Publ. Corporation,
New York; 1965
1,5

923 Hukill, W.V.
Grain drying.
In: Christensen, C.M (Ed.).
Storage of Cereal Grains and their Products. pp 481-508.
American Association of Cereal Chemists, Incorporated,
St. Paul, Minnesota; 1974
2,5

924 Humphries, W.R.; Hurst, W.M.
Moisture changes in some agricultural products due to
atmospheric conditions.
Agricultural Engineering 16 (1) 8-11; 1935
2,5

925 Hunt, D.G.
Prediction of sorption and diffusion of water vapor by
nylon 6,6.
Polymers 21 (-) 495-501; 1980
2,5

926 Hunt, W.H.; Neustadt, M.H.
Factors affecting the precision of moisture determination in
grain and related crops.
Journal of the Association of Official Analytical Chemists
49 (4) 757-763; 1966
3

927 Hunt, W.H.; Pixton, S.W.
Moisture - its significance, behavior, and measurement.
In: Christensen, C.M. (Ed.).
Storage of cereal grains and their products. pp 1-55.
American Association of Cereal Chemists, Incorporated,
St. Paul, Minnesota; 1974
1,3,4,5

928 Hunter, J.H.
Growth and aflatoxin production in shelled corn by the
Aspergillus flavus group as related to relative humidity and
temperature.
Ph.D. Thesis, Purdue University; 1969
4

929 Hunziker, M.
Inbetriebnahme einer neuartigen isopiestischen
Sorptionsapparatur und Bestimmung der Isopsychren von
Casein-Natriumchlorid-Gemischen bei hohen Wasseraktivitäten.
Inauguraldissertation, Universität Bern; 1973
3,5

930 Hunziker, O.F.; Nissen, B.H.
Lactose solubility and lactose crystal formation.
I. Lactose solubility.
Journal of Dairy Science 9 (-) 517-537; 1926
4

931 Hutchinson, J.B.; Martin, H.F.
The measurements of lipase activity in oat products.
Journal of the Science of Food and Agriculture 3 (-) 312; 1952
4

932 Huzayyin, A.S.; Hodges, T.O.; Miller, P.L.
Unsteady-state cooling of high moisture corn.
Transaction of the American Society of Agricultural
Engineers 717-723; 1973
2,5

933 Hwang, I.Y ; Park, K.H.
A study for the manufacture of starch based intermediate
moisture food.
Korean Journal of Food Science and Technology
13 (3) 227-232; 1981
4

934 Ida, M.; Sakabe, Y.; Nishimura, T.
The changes in dielectric properties of water during
adsorption by solids. II.
The Science Reports of Kanazawa University 13 (2) 91-99; 1968
3

935 Igbeka, J.C.; Blaisdell, J.L.
Moisture isotherms of a processed meat product - bologna.
Journal of Food Technology 17 (1) 37-46; 1982
2,3,5

936 Iglesias, H.A.
Isotermas de sorción de agua en remolacha azucarera y análisis
del fenómeno de sorción en alimentos.
Thesis, Dempartamento de Industrias, Facultad de Ciencias
Exactas y Naturales, Universidad de Buenos Aires, Argentina;
1975
1,2,5

937 Iglesias, H.A.; Boquet, R.; Chirife, J.
On the evaluation of B.E.T. constants from the B.E.T. isotherm
equation.
Journal of Food Science 42 (5) 1387-1389; 1977
2

938 Iglesias, H.A.; Chirife, J.
B.E.T. monolayer values in dehydrated foods and food
components.
Lebensmittel-Wissenschaft und -Technologie 9 (2) 107-113; 1976
2,5

939 Iglesias, H.A.; Chirife, J.
Isosteric heats of water vapor sorption on dehydrated foods.
I. Analysis of the differential heat curves.
Lebensmittel-Wissenschaft und -Technologie 9 (2) 116-122; 1976
2

940 Iglesias, H.A.; Chirife, J.
Isosteric heats of water vapor sorption on dehydrated foods.
II. Hysteresis and heat of sorption comparison with B.E.T.
theory.
Lebensmittel-Wissenschaft und -Technologie 9 (2) 123-127; 1976
2

941 Iglesias, H.A.; Chirife, J.
Prediction of the effect of temperature on water sorption
isotherms of food material.
Journal of Food Technology 11 (2) 109-116; 1976
2,5

942 Iglesias, H.A.; Chirife, J.
Equilibrium moisture contents of air dried beef. Dependence on
drying temperature.
Journal of Food Technology 11 (6) 565-573; 1976
2,5

943 Iglesias, H.A.; Chirife, J.
A model for describing the water sorption behavior of foods.
Journal of Food Science 41 (5) 984-992; 1976
2

944 Iglesias, H.A.; Chirife, J.
On the local isotherm concept and modes of moisture binding in
food products.
Journal of Agricultural and Food Chemistry 24 (1) 77-79; 1976
1,2

945 Iglesias, H.A.; Chirife, J.
Effect of fat content on the water sorption isotherm of air
dried minced beef.
Lebensmittel-Wissenschaft und -Technologie 10 (3) 151-152;
1977
5

946 Iglesias, H.A.; Chirife, J.
Effect of heating in the dried state on the moisture sorption
isotherm of beef.
Lebensmittel-Wissenschaft und -Technologie 10 (5) 249-250;
1977
5

947 Iglesias, H.A.; Chirife, J.
An empirical equation for fitting water sorption isotherms of
fruits and related products.
Canadian Institute of Food Science and Technology Journal
11 (1) 12-15; 1978
2,5

948 Iglesias, H.A.; Chirife, J.
Delayed crystallization of amorphous sucrose in humidified
freeze dried model systems.
Journal of Food Technology 13 (2) 137-144; 1978
2

949 Iglesias, H.A.; Chirife, J.
An equation for fitting uncommon water sorption isotherms in
foods.
Lebensmittel-Wissenschaft und -Technologie 14 (2) 105-106;
1981
2

Iglesias, H.A.; Chirife,J. see also 2201

950 Iglesias, H.A.; Chirife, J.; Boquet, R.
Prediction of water sorption isotherms of food models from
knowledge of components sorption behavior.
Journal of Food Science 45 (3) 450-452, 457; 1980
1,2,5

951 Iglesias, H.A.; Chirife, J.; Lombardi, J.L.
     An equation for correlating equilibrium moisture content in
     foods.
     Journal of Food Technology 10 (3) 289-297; 1975
     2,5
952 Iglesias, H.A.; Chirife, J.; Lombardi, J.L.
     Comparison of water vapor sorption by sugar beet root
     components.
     Journal of Food Technology 10 (4) 385-391; 1975
     1,5
953 Iglesias, H.A.; Chirife, J.; Lombardi, J.L.
     Water sorption isotherms in sugar beet root.
     Journal of Food Technology 10 (3) 299-308; 1975
     2,5
954 Iglesias, H.A.; Chirife, J.; Viollaz, P.
     Thermodynamics of water vapor sorption by sugar beet root.
     Journal of Food Technology 11 (1) 91-101; 1976
     2,5
955 Iglesias, H.A.; Chirife, J.; Viollaz, P.
     Evaluation of some factors useful for the mathematical
     prediction of moisture gain by packaged dried beef.
     Journal of Food Technology 12 (5) 505-513; 1977
     2
956 Iglesias, H.A.; Viollaz, P.; Chirife, J.
     A technique for predicting moisture transfer in mixtures of
     packaged dehydrated foods.
     Journal of Food Technology 14 (1) 89-93; 1979
     2,5
957 Ingram, M.
     Microorganisms resisting high concentrations of sugar or
     salts.
     VII. Symposium of the Society for General Microbiology
     7 (-) 90-133; 1957
     4
958 Insalata, N.F.
     The technical microbiological problems in intermediate
     moisture products.
     Food Product Development 6 (5) 72, 74, 76; 1972
     4
959 Irving, C.L.; Higgins, C.T.
     HVAC system narrows relative humidity control to +- 0.1%.
     Control Engineering 27 (11) 150-158; 1980
     3
960 Ito, K.; Kaga, S.; Takeya, Y.
     Studies on hard gelatin capsules.
     I. Water vapor transfer between capsules and powders.
     Chem. Pharm. Bull. 17 (6) 1134-1137; 1969
     5
961 Ives, N.C.
     A dew point moisture indicator.
     Journal of the Agricultural Engineering, St. Joseph Michigan
     33 (2) 85; 1952
     3
962 Iyengar, J.R.; Sen, D.P.
     The equilibrium relative humidity relationship of salted fish
     (Barbus carnaticus and Rastrelliger canagurta): The effect of
     calcium and magnesium as impurities in common salt used for
     curing.
     Journal of Food Science and Technology (Mysore) 7 (1) 17-19;
     1970
     5

963 Jakobsen, M.
A comparative study of methods for water activity
measurement.(Orig.dan.)
Nordisk Veterinaermedicin 30 (10) 437-450; 1978
3

964 Jakobsen, M.; Filtenborg, O.
Measurement of water activity of meat products. Collaborative
trials with the "aw Meter Modell 5803".
Dansk Veterinaertidsskrift 63 (4) 144-150; 1980
3

965 Jakobsen, M.; Filtenborg, O.; Bramsnaes, F.
Germination and outgrowth of the bacterial spore in the
presence of different solutes.
Lebensmittel-Wissenschaft und -Technologie 5 (5) 159-162; 1972
4

966 Jakobsen, M.; Jensen, H.C.
Combined effect of water acitivity and pH on the growth of
butyric anaerobes in canned pears.
Lebensmittel-Wissenschaft und -Technologie 8 (4) 158-160; 1975
4

967 Jakobsen, M.; Trolle, G.
The effect of water activity on growth of clostridia.
Nordisk Veterinaermedicin 31 (5) 206-213; 1979
4

968 Janis, A.A.; Ferguson, J.B.
Sodium chloride solutions as an isopiestic standard.
Canadian Journal of Research (B) 17 (-) 215-230; 1939
3

969 Jarvis, B.
Do mycotoxins present a potential hazard for intermediate
moisture foods?
In: Davies, R.; Birch,. G.G.; Parker, K.J. (Edts.).
Intermediate Moisture Foods. pp 239-247.
Applied Science Publ., London; 1976
4

970 Jason, A.C.
A study of evaporation and diffusion processes in the drying
of fish muscle.
In: Fundamental aspects of the dehydration of foodstuffs.
pp 103-135.
Conference in Aberdeen 1958.Society of Chemical Industry,
London; 1958
2,5

971 Jason, A.C.
Relative humidity in food industry; Introduction
British Food Manufacturing Industries Research Association
Symposium Proc. No. 4, pp 1 and 44-51. Leatherhead,
London; 1969
1

972 Jason, A.C.
A note on the systematic errors in water activity measurements
of MCC caused by radiative heat loss.
In: Jowitt, R.; Escher, F.; Hallström, B.; Meffert, H.F.T.;
Spieß, W.E.L.; Vos, G. (Edts.).
Physical Properties of Foods. pp 88-91.
Applied Science Publ., London, New York; 1983
3

973 Jay, E.G.; Arbogast, R.T.; Pearman, G.C.
Relative humidity: Its importance in the control of
stored-product insects with modified atmospheric gas
concentrations.
Journal of Stored Products Research 6 (-) 325-329; 1971
4

974 Jayaraman, K.S.; Dasgupta, D.K.
     Development and storage stability of intermediate moisture
     carrot.
     Journal of Food Science 43 (-) 1880-1881; 1978
     4
975 Jayaraman, K.S.; Ramanuja, M.N.; Goverdhanan, T.;
     Bhatia, B.S.; Nath, H.
     Technological aspects of use of ripe mangoes in the
     preparation of convenience foods for defence services.
     Indian Food Packer 30 (5) 76-82; 1976
     4
976 Jayaraman, K.S.; Ramanuja, M.N.; Nath, H.
     A modified graphical interpolation method for rapid
     determination of water activity in foods.
     Journal of Food Science and Technology 14 (-) 129-130; 1977
     3,5
977 Jayaratnam, S.; Kirtisinghe, D.
     The effect of relative humidity and temperature on moisture
     sorption by black tea.
     Tea Quarterly 44 (4) 164-169; 1974
     5
978 Jayaratnam, S.; Kirtisinghe, D.
     The effect of relative humidity on the storage life of made
     tea.
     Tea Quarterly 44 (4) 170-172; 1974
     4,5
979 Jeffries, R.
     Sorption of water by cellulose.
     Methods in carbohydrate chemistry .
     In: Whistler (Ed.). Vol. III, Cellulose. pp 120-127.
     Academic Press, New York; 1963
     3,5
980 Jelaca, S.L.; Hlynka, I.
     Water binding capacity of wheat flour crude pentosans and
     their relation to mixing characteristics of dough.
     Cereal Chemistry 48 (-) 211-222; 1971
     2
981 Jericevic, D.; Le Maguer, M.
     Influence of the moisture content on the rate of rehydration
     of potato granules.
     Journal de l'Institut Canadien de Technologie Alimentaire
     8 (2) 88-91; 1975
     4,5
982 Jermolenko, W.D.
     Untersuchungen der Bindungsform des Wassers bei
     Lebensmitteln durch Methoden der physikalischen Chemie.
     Die Stärke 14 (1) 30; 1962
     2
983 Jodl, R.
     Untersuchungen über die Wasserbindungsverhältnisse des
     Lignins. Eine Modellstudie als Beitrag zur Klärung der
     Wasserverbindungsverhältnisse des Tabaks.
     Beiträge zur Tabakforschung 4 (7) 309-312; 1968
     2,5
984 John, P.T.; Sekhorn, R.S.
     Methods of measuring vapor pressure of water and other
     liquids.
     Journal of Scientific and Industrial Research 27 (-) 50-57;
     1968
     3

985 Johnson, C.G.
The maintenance of high atmospheric humidities for
entomological work with glycerol water mixtures.
Annals of Applied Biology 27 (-) 295-299; 1940
3
986 Johnson, C.M.
Determination of water in dry food materials - application of
the Fischer volumetric method.
Industrial and Engineering Chemistry, Analytical Edition
17 (5) 312-316 ; 1945
3
987 Johnson, E.F.; Molstad, M.C.
Thermodynamic properties of aqueous lithium chloride
solutions.
The Journal of Physical and Colloid Chemistry 55 (-) 257-281;
1951
2,3
988 Johnson, J.L.; Busk, G.C.; Labuza, T.P.
Examination of the cristallinity of food gels by X-ray
diffraction.
Journal of Food Science 45 (-) 77-83; 1980
4
989 Johnson, R.G.; Sullivan, D.B.; Secrist, J.L.; Brockmann, M.C.
The effect of high temperature storage on the acceptability
and stability of intermediate moisture food.
Natick Laboratories Technical Report 72-76-FL
4
990 Jokisch, F.
Über den Stofftransport im hygroskopischen Feuchtebereich
kapillarporöser Stoffe am Beispiel des Wasserdampftransports
in technischen Adsorbentien.
Dissertation, Technische Hochschule Darmstadt; 1975
1,2,5
991 Jokisch, F.; Kast, W.
Transportmechanismen im hygroskopischen Feuchtebereich poröser
Stoffe.
Chemie-Ingenieur-Technik 47 (9) 393; 1975
1,2
992 Jones, D.C.
Some comments on the B.E.T. (Brunauer-Emmett-Teller) adsorption
equation.
Journal of the Chemical Society (-) 126-130; 1951
2
993 Jones, F.R.
An accurate method for the determination of aqueous vapour
pressure: the equilibrium humidities of solutions of sulphuric
acid.
Journal of Applied Chemistry 1 (2) 144-152; 1951
3
994 Jones, N.R.
The meaning and importance of relative humidity.
In: Relative humidity in the food industry.
British Food Manufacturing Industries Research Association
Symposium Proc. No. 4, pp 2-8, Leatherhead, London; 1969
1,5
995 Jordao, B.A.; Stolf, S.R.
Curva de saturacao do feijao de mesa variedade rosinha.
Coletanea do Instituto de Tecnologia de Alimentos Campinas,
Brazil 3 (-) 425; 1970
5

996 Joslyn, M.A.
The action of enzymes in concentrated solution and in the
dried state.
Journal of the Science of Food and Agriculture 2 (-) 289-294;
1951
4

997 Joyner, L.G.
Moisture control in sugar.
Modern Packaging 29 (12) 180-184, 238-240, 242, 244; 1956
2,3,5

998 Jüntgen, H.; Knoblauch, K.; Richter, E.
Grundlagen der Adsorption.
Fortschritte der Verfahrenstechnik 17 (-) 247-266; 1979
1

999 Juliano, B.O.
Hygroscopic equilibrium of rough rice.
Cereal Chemistry 41 (5) 191-197; 1964
5

1000 Jura, G.; Harkins, W.D.
A new adsorption isotherm which is valid over a very wide
range of pressure.
Journal of Chemical Physics 11 (9) 430-431; 1943
2

1001 Kacenak, I.
Sorption-desorption properties of bread and their relation to
packaging.(Orig.czech.)
Mlynsko-Pekarensky Prumysl 24 (6) 186-192; 1978
4,5

1002 Kacenak, I.
Sorptionseigenschaften von Brot und ihre Auswirkung beim
Verpacken.
Bäcker und Konditor 27 (12) 354-357; 1979
4,5

1003 Kachru, R.P.; Matthes, R.K.
The behaviour of rough rice in sorption.
Journal of Agricultural Engineering Research 21 (-) 405-416;
1976
4,5

1004 Kadlec, O.; Dubinin, M.M.
Comments on the limits of applicability of the mechanism of
capillary condensation.
Journal of Colloid Interface Science 31 (4) 479-489; 1969
2,5

1005 Kaess, G.
Über die Haltbarkeit und Verpackung einiger Süßwaren.
Deutsche Lebensmittel Rundschau 45 (2) 29-40; 1949
2,3,4,5

1006 Kahn, M.L.; Eapen, K.E.
Intermediate-moisture frozen foods.
United States Patent 4 244 976; 1981
4

1007 Kamman, J.F.; Labuza, T.P.; Warthesen, J.J.
Kinetics of thiamin and riboflavin loss in pasta as a function
of constant and variable storage conditions.
Journal of Food Science 46 (5) 1457-1461; 1981
4

1008 Kamman, J.F.; Tatini, S.R.; Labuza, T.P.
Effect of water activity on nuclease production by
Staphylococcus aureus.
Journal of Food Science 43 (4) 1284-1286, 1292; 1978
4

1009 Kammerer, F.X.
Zuckeraustauschstoffe bei der Süßwaren-Herstellung.
Kakao und Zucker 24 (5) 184-190; 1972
5

1010 Kanagy, J.R.
Adsorption of water vapor by untanned hide and various
leathers at 100°F.
Journal of Research of the National Bureau of Standards
38 (-) 119-125; 1947
2,3,5

1011 Kanagy, J.R.; Cassel, J.M.
Heats of wetting of modified collagen and other materials.
The Journal of American Leather Chemist's Association
52 (-) 248-259; 1957
2,5

1012 Kang, A.S.; Schornick, G.; Loncin, M.
Aromaerhaltung beim Sprühtrocknen von Zuckerlösungen.
Zeitschrift für Lebensmitteltechnologie und -Verfahrenstechnik
30 (7) 317-320; 1979
2

1013 Kang, C.K.; Woodburn, M.; Pagenkopf, A.; Cheney, R.
Growth, sporulation and germination of Clostridium perfringens
in media of controlled water activity.
Applied Microbiology 18 (-) 798-805; 1969
4

1014 Kanner, J.; Mendel, H.; Budowski, P.
Carotene oxidizing factors in red pepper fruits (Capsicum
annuum L.): Oleoresin-cellulose solid model.
Journal of Food Science 43 (-) 709-712; 1978
4,5

1015 Kaplow, M.
Commercial development of intermediate moisture foods.
Food Technology 24 (8) 889-893; 1970
1,4,5

1016 Kapsalis, J.G.
Hygroscopic equilibrium and texture of freeze-dried foods.
United States Army, Natick Laboratories, Natick Technical
Report 67-87-FL; 1967
1,2,4,5

1017 Kapsalis, J.G.
Moisture and food characteristics.
In: The Skylab program.
USA, Research Development Associates for Military Food &
Packaging Systems Inc.
Activities Report 25 (1) 60-69; 1973
4

1018 Kapsalis, J.G.
The influence of water on textural parameters in foods at
intermediate moisture levels.
In: Duckworth, R.B. (Ed.).
Water Relations of Foods. pp 627-637.
Academic Press, London; 1975
4

1019 Kapsalis, J.G.
Moisture sorption hysteresis.
In: Rockland, L.B.; Stewart, G.F. (Edts.).
Water Activity: Influences on Foodquality. pp 143-177.
Academic Press, New York; 1981
2

1020   Kapsalis, J.G.; Drake, B.; Johansson, B.
        Textural properties of dehydrated foods. Relationships with
        the thermodynamics of water vapor sorption.
        Journal of Texture Studies 1 (3) 285-308; 1970
        1,2,4
1021   Kapsalis, J.G.; Segars, R.
        Rheological properties of meats (fresh, intermediate moisture
        and dehydrated).
        Lebensmittel-Wissenschaft und -Technologie 9 (6) 383-385; 197
        4
1022   Kapsalis, J.G.; Walker, J.E. jr.; Wolf, M.
        A physico-chemical study of the mechanical properties of low
        and intermediate moisture foods.
        Journal of Texture Studies 1 (4) 464-483; 1970
        1,2,4
1023   Kapsalis, J.G.; Wolf, M.; Driver, M.; Henick, A.S.
        The moisture sorption isotherm as a basis for the study of
        sorption and stability characteristics in dehydrated foods.
        Proc. 16. Research Conference, Research Council
        of the American Meat Institute Foundation. 73-93; 1964
        1,2,3,4
1024   Kapsalis, J.G.; Wolf, M.; Driver, M.; Henick, A.S.
        Humidity and moisture measurements in relation to storage
        stability of dehydrated foods.
        In: Wexler, A.; Amdur, E.J. (Edts.).
        Humidity and Moisture. Measurement and Control in Science and
        Industry. Vol. 2, pp 161-172. Reinhold Publ. Corporation,
        New York; 1965
        3,4
1025   Kapsalis, J.G.; Wolf, M.; Driver, M.; Walker, J.W.
        The effect of moisture content on the flavor and texture
        stability of dehydrated foods.
        ASHRAE Journal 13 (-) 93-99; 1971
        4
1026   Karan-Djurdjic, S.; Leistner, L.
        Messung der Wasseraktivität von Fleisch und Fleischwaren mit
        dem SINA-Gerät.
        Die Fleischwirtschaft 50 (8) 1104-1106; 1970
        3
1027   Karel, M.
        Recent research and development in the field of low-moisture
        and intermediate-moisture foods.
        Critical Reviews in Food Technology 3 (3) 329-372; 1973
        1
1028   Karel, M.
        Stability of low and intermediate moisture foods.
        In: Goldblith, S.A.; Rey, L.; Rothmayr, W.W. (Edts.).
        Freeze Drying and Advanced Food Technology. pp 643-674.
        Academic Press, London; 1975
        1,4
1029   Karel, M.
        Free radicals in low moisture systems.
        In: Duckworth, R.B. (Ed.).
        Water Relations of Foods. pp 435-453.
        Academic Press, London; 1975
        4
1030   Karel, M.
        Physico-chemical modification on the state of water in foods
        - a speculative survey.
        In: Duckworth, R.B. (Ed.).
        Water Relations of Foods. pp 639-657.
        Academic Press, London; 1975
        1,2

1031 Karel, M.
      Water activity and food preservation.
      In: Fennema, O. (Ed.). Principles of Food Science. Part II.
      Physical Principles of Food Preservation. pp 237-263.
      Marcel Deccer, New York, Basel; 1975
      1,3,4
1032 Karel, M.
      Technology and application of new intermediate moisture foods.
      In: Davies, R.; Birch, G.G.; Parker, K.J. (Edts.).
      Intermediate Moisture Foods. pp 4-31.
      Applied Science Publ., London; 1976
      4
1033 Karel, M.; Aikawa, Y.; Proctor, B.E.
      New approach to humiditiy equilibria data.
      Modern Packaging 29 (2) 153-156, 237-240; 1955
      2,3,5
1034 Karel, M.; Labuza, T.P.
      Nonenzymatic browning in model systems containing sucrose.
      Journal of Agricultural and Food Chemistry 16 (5) 717-719;
      1968
      4
1035 Karel, M.; Mizrahi, S.; Labuza, T.P.
      Computer prediction of food storage.
      Modern Packaging 44 (8) 54-58; 1971
      4
1036 Karel, M.; Nickerson, J.T.R.
      Effects of relative humidity, air and vacuum in browning of
      dehydrated orange juice.
      Food Technology 18 (8) 1214-1218; 1964
      3,4,5
1037 Karel, M.; Yong, S.
      Autoxidation - initiated reactions in foods.
      In: Rockland, L.B.; Stewart, G.F. (Edts.).
      Water Activity: Influences on Foodquality. pp 511-529.
      Academic Press, New York; 1981
      4
1038 Kargin, V.A.
      Sorption properties of glasslike polymers.
      Journal of Polymer Science 23 (-) 47-55; 1957
      2,5
1039 Karmas, E.
      Water in biosystems.
      Journal of Food Science 38 (5) 736-739; 1973
      2
1040 Karmas, E.
      Die Bedeutung der Wasseraktivität in gefrorenen Lebensmitteln.
      Temperatur Technik 15 (5) 141-143; 1977
      1,4
1041 Karmas, E.
      Techniques for measurement of moisture content of foods.
      Food Technology 34 (4) 52-59; 1980
      3
1042 Karmas, E.; Chen, C.C.
      Relationship between water activity and water binding in high
      and intermediate moisture foods.
      Journal of Food Science 40 (4) 800-801; 1975
      2,4
1043 Karon, M.L.
      Hygroscopic equilibrium of cottonseed.
      Journal of the American Oil Chemists Society 24 (2) 56-58;
      1947
      5

1044 Karon, M.L.; Adams, M.E.
Note on the hygroscopic equilibrium of cottonseed and
cottonseed products.
Journal of the American Oil Chemists Society 25 (1) 21-22;
1948
5

1045 Karon, M.L.; Adams, M.E.
Hygroscopic equilibria of rice and ricefractions.
Cereal Chemistry 26 (1) 1-12; 1949
5

1046 Karon, M.L.; Hillery, B.E.
Hygroscopic equilibrium of peanuts.
Journal of the American Oil Chemists Society 26 (1)
1949
5

1047 Kashurin, A.N.; Domaretskii, V.A.
Hydrothermic equilibrium in barley malt during desorption.
(Orig.russ.)
Izvestiya Vysshikh Uchebnykh Zavedenii, Pishchevaya
Tekhnologiya (5) 125-127; 1976
2,3,5

1048 Kaspers, W.
Relative und Gleichgewichtsfeuchte in der Lebensmittel-
technologie.
Süßwaren 26 (6) 274-276; 1981
1

1049 Kast, W.
Adsorption aus der Gasphase - Grundlagen und Verfahren.
Chemie-Ingenieur-Technik 53 (3) 160-172; 1981
1

1050 Kast, W.; Jokisch, F.
Überlegungen zum Verlauf von Sorptionsisothermen und zur
Sorptionskinetik an porösen Feststoffen.
Chemie-Ingenieur-Technik 44 (8) 556-563; 1972
2,3

1051 Katchalski, E.
Poly-alpha-amino acids.
Advances in Protein Chemistry 6 (-) 123-185; 1951
5

1052 Katchman, B.; McLaren, A.D.
Sorption of water vapors by proteins and polymers. IV.
Journal of the American Chemical Society 73 (-) 2124-2127;
1951
1,2

1053 Katz, J.R.
Ph.D. Thesis, University of Amsterdam; 1917
5

1054 Katz, J.R.
Die Gesetze der Quellung.
Kolloidchemische Beihefte 9 (-) 6; 1917
5

1055 Katz, M.J.
An explicit function for specific surface area.
Journal of Analytical Chemistry 26 (4) 734-735; 1954
3

1056 Katz, E.E.; Labuza, T.P.
Effect of water activity on the sensory crispness and
mechanical deformation of snack food products.
Journal of Food Science 46 (2) 403-409; 1981
4,5

1057 Kaufmann, H.P.; Thieme, J.G.
Neuzeitliche Technologie der Fette und Fettprodukte.
XXII: Die Lagerung der Rohstoffe.
Fette, Seifen, Anstrichmittel 58 (2) 131-138; 1956
4,5

1058 Kawaguchi, Y.; Kanai, H.; Kajiwara, H.; Arai, Y.
Correlation for activities of water in aqueous electrolyte
solutions using ASOG model.
Journal of Chemical Engineering, Japan 14 (3) 243-246; 1981
3

1059 Kawasaki, K.; Kanou, K.
Control of atmospheric humidity by aqueous sulfuric acid
solution.
In: Wexler, A.; Wildhack, W.A. (Edts.).
Humidity and Moisture. Measurement and Control in Science and
Industry. Vol. 3, pp 531-534. Reinhold Publ. Corporation,
New York; 1965
3

1060 Kearsley, M.W.
The control of hygroscopicity, browning and fermentation in
glucose syrups.
Journal of Food Technology 13 (-) 339-348; 1978
4

1061 Kearsley, M.W.; Birch, G.G.
Selected physical properties of glucose syrup fractions
obtained by reverse osmosis.
II. Hygroscopicity.
Journal of Food Technology 10 (-) 625-635; 1975
3,5

1062 Kelly, J.F.; Fuller, O.M.
An evaluation of a method for investigating sorption and
diffusion in porous solids.
Industrial and Engineering Chemistry. Fundamentals
19 (1) 11-17; 1980
2

1063 Kelsey, K.E.
The sorption of water vapour by wood.
Australian Journal of Applied Science 8 (1) 42-54; 1957
2,3,5

1064 Kennerley, M.G.
A technique for the measurement of the water adsorption of
small amounts of hygroscopic materials. Water adsorption
isotherm of polyglycine.
Polymer 10 (-) 833-839; 1969
3,5

1065 Kent, M.
Complex permittivity of white fish meal in the microwave
region as a function of temperature and moisture content.
Journal of Physics. D, Applied Physics 3 (-) 1275-1283; 1970
2

1066 Kent, M.
Measurement of the hydration of biological molecules from
dielectric measurements at centimetre wavelength.
In: Peeters, H. (Ed.). Proc. 19th Colloquium of the
Protides of Biological Fluids; Brugge, Belgium; 1971
3

1067 Kent, M.
Microwave dielectric properties of fishmeal.
Journal of Microwave Power 7 (2) 109-116; 1972
2,3

1068 Kent, M.
      Complex permittivity of protein powders at 9.4 GHz as a
      function of temperature and hydration.
      Journal of Physics D, Applied Physics 5 (-) 394-409; 1972
      3,5
1069 Kent, M.; Jason, A.C.
      Dielectric properties of foods in relation to interactions
      between water and substrate.
      In: Duckworth, R.B. (Ed.). Water Relations of Food. pp 211-231
      Academic Press, London; 1975
      3
1070 Kent, M.; Meyer, W.
      Dielectric relaxation of adsorbed water in microcrystalline
      cellulose.
      Journal of Physics D, Applied Physics 16 (-) 915-925; 1983
      2,3
1071 Kertesz, Z.
      Water relations of enzymes. 2. Water concentration required
      for invertase action.
      Journal of the American Chemical Society 57 (-) 1277-1279;
      1935
      4
1072 Kertesz, Z.I.; Massey, L.M. jr.; Parsons, G.F.; Simon, M.
      Storage behaviour of powdered dehydrated cranberries.
      Food Technology 17 (12) 1569-1572; 1963
      4
1073 Kessler, H.G.
      Probleme der Lebensmitteltrocknung.
      Chemie-Ingenieur-Technik 47 (18) 755-759; 1975
      1,2
1074 Kidambi, R.N.; Wiebe, H.H.; Ernstrom, C.A.; Richardson, G.H.
      An economical dewpoint instrument for rapid measurement of
      water activity in foods.
      Journal of Dairy Science 62 (Suppl. 1) 40-41; 1979
      3
1075 Kiermeier, F.; Coduro, E.
      Über den diastatischen Stärkeabbau in lufttrockenen
      Substanzen.
      Biochemische Zeitschrift 325 (-) 280-287; 1954
      4
1076 Kiermeier, F.; Coduro, E.
      Der Einfluß des Wassergehaltes auf Enzymreaktionen in
      wasserarmen Lebensmitteln.
      Zeitschrift für Lebensmittel-Untersuchung und -Forschung
      98 (-) 119-129; 1954
      4,5
1077 Kiermeier, F.; Coduro, E.
      Der Einfluß des Wassergehaltes auf Enzymreaktionen in
      wasserarmen Lebensmitteln.
      III. Mitteilung: Über Enzymreaktionen in Mehl, Teig und Brot.
      Zeitschrift für Lebensmittel-Untersuchung und -Forschung
      102 (1) 7-12; 1955
      4
1078 Kilara, A.; Humbert, E.S.; Sosulski, F.W.
      Nitrogen extractability and moisture adsorption
      characteristics of sunflower seed products.
      Journal of Food Science 37 (5) 771-773; 1972
      5
1079 Kim, M.N.; Saltmarch, M.; Labuza, T.P.
      Non-enzymatic browning of hygroscopic whey powders in open
      versus sealed pouches.
      Journal of Food Processing and Preservation 5 (1) 49-57; 1981
      4,5

1080 King, A.D. jr.; Halbrook, W.U.; Fuller, G.; Whitehand, L.C.
     Almond nutmeat moisture and water activity and its influence
     on fungal flora and seed composition.
     Journal of Food Science 48 (2) 615-617; 1983
     2,4,5
1081 King, C.J.
     Rates of moisture sorption and desorption in porous, dried
     foodstuffs.
     Food Technology 22 (4) 165-171; 1968
     2
1082 King, C.J.; Carn, R.M.; Jones, R.L.
     Processing approaches for limited freeze drying.
     Journal of Food Science 41(3) 612-618; 1976
     2
1083 King, J.C.; Lam, W.K.; Sandall, O.C.
     Physical properties important for freeze-drying poultry meat.
     Food Technology 22 (10) 1302-1308; 1968
     3,5
1084 Kiozumi, C.; Wada, S.; Nanaka, L.
     A modified graphic interpolation method for measurements of
     wateractivity and effect of ingradient on wateractivity.
     Journal of the Tokyo University of Fisheries (Japan)
     67 (1) 29-34; 1980
     3
1085 Kirchmeier, O.
     Der besondere Lösungszustand von Caseinatschmelzen.
     Zeitschrift für Lebensmittel-Untersuchung und -Forschung
     166 (5) 293-297; 1978
     5
1086 Kirievskii, B.N.
     Sorption hysteresis of sunflower seeds.(Orig.russ.)
     Trudy Vsesoyuznogo Nauchno-Issledovatel'skogo Instituta Zhirov
     27 (-) 142-145; 1970
     2,5
1087 Kirk, J.R.
     Influence of water activity on stability of vitamins in
     dehydrated foods.
     In: Rockland, L.B.; Stewart, G.F. (Edts.).
     Wateractivity: Influences on Foodquality. pp 531-566.
     Academic Press, New York; 1981
     4
1088 Kirk, J.R.; Dennison, D.B.; Kokoczka, P.; Heldman, D.R.
     Degradation of ascorbic acid in a dehydrated food system.
     Journal of Food Science 42 (5) 1274-1279; 1977
     4
1089 Kiss, I.; Farkas, J.
     The storage of wheat and corn of high moisture content as
     affected by ionizing radiation.
     Acta Alimentaria 6 (3) 193-214; 1977
     4
1090 Kisselew, A.W.
     Spezifische Oberfläche von Adsorbenzien verschiedener
     Struktur. Absolute Isothermen und Adsorptionswärmen.
     In: Witzmann, H. (Ed.). Methoden der Strukturuntersuchung an
     hochdispersen und porösen Stoffen.
     Akademieverlag, Berlin; 1961
     2
1091 Kleint, C.; Brzoska, K.D.
     Untersuchungen von Sorptionsphänomenen mittels
     Desorptionsspektrometrie (Grundlagen, Anwendungen, Probleme)
     Beilage zur Zeitschrift "Vakuum Information" der VEB
     Hochvakuum Dresden. DDR; 1972
     1

1092 Klettner, P.G.; Rödel, W.
Überprüfung und Steuerung wichtiger Parameter bei der
Rohwurstreifung.
Die Fleischwirtschaft 58 (1) 57-66; 1978
4

1093 Kluge, G.; Heiss, R.
Untersuchungen zur besseren Beherrschung der Qualität von
getrockneten Lebensmitteln unter besonderer Berücksichtigung
der Gefriertrocknung.
Verfahrenstechnik 1 (6) 251-260; 1967
2,5

1094 Kneule, F.
Sorptionsisothermen.
Berichtsheft 6 der Fachgemeinschaft Allgemeine Lufttechnik im
VDMA (Verein Deutscher Maschinenbau-Anstalten) Frankfurt/M.;
1964
5

1095 Kobayashi, J.; Toyama, Y.
The aging effect of an electrolytic hygrometer.
In: Wexler, A.; Ruskin, R.E. (Edts.).
Humidity and Moisture. Measurement and Control in Science and
Industry. Vol. 1, pp 248-264. Reinhold Publ. Corporation,
New York; 1965
3

1096 Koehler, B.
Fungus growth on shelled corn as affected by moisture.
Journal of the Agricultural Research 56 (-) 291; 1938
4

1097 Köhler, W.; Riedel, O.; Scherer, H.
Ein mit Heizimpulsen gesteuertes isothermes Kalorimeter.
II. Anwendungsbeispiele für kalorische und
kalorisch-kinetische Messungen.
Chemie-Ingenieur-Technik 45 (22) 1289-1294; 1973
2

1098 Koga, S.; Maeda, Y.
Thermal properties of water in relation to microbial cells.
In: Rockland, L.B.; Stewart, G.F. (Edts.).
Water Activity: Influences on Foodquality. pp 813-823.
Academic Press, New York; 1981
2,4

1099 Koiwa, Y.; Ohta, S.
Lowering of water activity accompanying emulsion formation.
International Congress of Food Science & Technology -
Abstracts. p. 209; 1978
4

1100 Koizumi, C.; Iiyama, S.; Wada, S.; Nonaka, J.
Lipid deteriorations of freeze-dried fish meats at different
equilibrium relative humidities.
Bulletin of the Japan Society of Science Fisheries
44 (-) 209; 1978
5

1101 Kollmann, F.
Zur Theorie der Sorption.
Forschung auf dem Gebiete des Ingenieurwesens, Düsseldorf
29 (2) 33-41; 1963
2

1102 Kollmann, F.; Schneider, A.
Einrichtungen zur praxisnahen und wissenschaftlich exakten
Messung von Sorptionseigenschaften von Holz und
Holzwerkstoffen.
Holz als Roh- und Werkstoff 16 (4) 117-122; 1958
3

1103 Komeyasu, M.; Iyama, M.
Studies on spray-dried citrus unshiu juice.
Part 1. Characteristics of water adsorption properties of
spray-dried citrus unshiu juice.(Orig.jap.)
Nippon Shokuhin Kogyo Gakkaishi 21 (-) 384; 1974
5

1104 Konstance, R.P.; Craig, J.G. Jr.; Panzer, C.C.
Moisture sorption isotherms for bacon slices.
Journal of Food Science 48 (1) 127-130; 1983
5

1105 Kopelman, I.J.; Meydav, S.; Weinberg, S.
Storage studies of freeze dried lemon crystals.
Journal of Food Technology 12 (4) 403-410; 1977
4,5

1106 Kopelman, I.J.; Saguy, I.
Drum dried beet powder.
Journal of Food Technology 12 (6) 615-621; 1977
2,5

1107 Korsch, D.
Das Lösungs- und Adsorptionsverhalten der Metalle Au, Pt, Pd
und Ru in Bezug auf die Gase H2, O2, N2, A und Wasserdampf.
Zeitschrift für Angewandte Physik 28 (2) 43-50; 1969
3,5

1108 Koury, B.J.; Spinelli, J.
Effect of moisture, carbohydrate and atmosphere on the
functional stability of fish protein isolates.
Journal of Food Science 40 (-) 58-61; 1975
4,5

1109 Kovalchuk, V.I.; Krivosheyev, Y.I.
Equilibrium moisture content and thermophysical
characteristics of sunflower seed.
Pitschewaja Technol. (1) 180-181; 1972
5

1110 Kovats, E.;Plattner, P.A.; Günthard, H.H.
Adsorptionsisothermen einiger Dehydrierungskatalysatoren.
Helvetica, Chimica Acta 37 (-) 997-1003; 1954
1,2,3

1111 Kracher, F.
Xylit: Bedeutung - Wirkung - Anwendung.
Kakao und Zucker 27 (3) 68-75; 1975,
27 (4) 108-112; 1975
5

1112 Kreisman, L.N.; Labuza, T.P.
Storage stability of intermediate moisture food process cheese
food products.
Journal of Food Science 43 (2) 341-344; 1978
4

1113 Krispien, K.
Versuche zur Ermittlung der Oberflächenwasseraktivität
(aw-Wert) von Schlachttierkörpern und Fleisch und zur
Beeinflussung der aw- und pH-Werte der Fleischoberfläche.
Dissertation, Freie Universität Berlin; 1978
1,3

1114 Krispien, K.; Rödel, W.
Bedeutung der Temperatur für den aw-Wert von Fleisch und
Fleischerzeugnissen.
Die Fleischwirtschaft 56 (5) 709-714; 1976
1,5

1115 Krispien, K.; Rödel, W.; Leistner, L.
Vorschlag zur Berechnung der Wasseraktivität (aw-Wert) von
Fleischerzeugnissen aus den Gehalten von Wasser und Kochsalz.
Die Fleischwirtschaft 59 (8) 1173-1177; 1979
2,5

1116 Kühn, I.
      A new theroretical analysis of adsorption phenomena.
      Introductory part: The characteristic expression of the main
      regular types of adsorption isotherms by a single simple
      equation.
      Journal of Colloid Science 19 (-) 685-698; 1964
      2
1117 Kulaba, G.W.; Henderson, S.
      Moisture content / equilibrium relative humidity relationship c
      parchment coffee.
      Kenya Coffee 45 (534) 271-276; 1980
      5
1118 Kumar, M.
      Moisture distribution between whole corn, endosperm and germ
      by various methods of conditioning.
      Journal of Food Technology 8 (4) 407-414; 1973
      2,5
1119 Kumar, M.
      Water vapour adsorption on whole corn flour, degermed corn
      flour, and germ flour.
      Journal of Food Technology 9 (4) 433-444; 1974
      2,5
1120 Kuntz, I.D.
      Hydration of macromolecules. III. Hydration of polypeptides.
      Journal of the American Chemical Society 93 (2) 514-516; 1971
      2
1121 Kuntz, I.D. jr.
      The physical properties of water associated with
      biomacromolecules.
      In: Duckworth, R.B. (Ed.). Water Relations of Foods. pp 93-109
      Academic Press, London; 1975
      2
1122 Kuntz, I.D.; Brassfield, T.S.; Law, G.D.; Purcell, G.V.
      Hydration of macromolecules.
      Science 163 (-) 1329-1331; 1969
      2
1123 Kuntz, I.D. jr.; Kauzmann, W.
      Die Hydratation von Proteinen und Polypeptiden.
      Advances in Protein Chemistry 28 (-) 239-342; 1974
      2,5
1124 Kunze, O.R.
      Moisture adsorption influences on rices.
      Journal of Food Process Engineering 1 (2) 167-181; 1977
      4
1125 Kunze, O.R.; Choudhury, M.S.U.
      Moisture adsorption related to the tensile strength of rice.
      Cereal Chemistry 49 (6) 684-696; 1972
      4
1126 Kunze, O.R.; Hall, C.W.
      Relative humidity changes that cause brown rice to crack.
      Transaction of the American Society of Agricultural Engineers
      8 (3) 396-399, 405; 1965
      2,4
1127 Kunze, O.R.; Hall, C.W.
      Moisture adsorption characteristics of brown rice.
      Transactions of the American Society of Agricultural Engineers
      10 (4) 448-450, 453; 1967
      2

1128   Kuppinger, H.; Müller, H.M.; Mühlbauer, W.
        Die Belüftungstrocknung von erntefrischem und vorgetrocknetem
        Körnermais unter thermodynamischem und mikrobiologischem
        Aspekt.
        Mitteilung aus dem Sonderforschungsbereich 140 -Landtechnik
        "Verfahrenstechnik der Körnerfruchtproduktion" der Universität
        Hohenheim
        Grundlagen der Landtechnik 27 (4) 119-128; 1977
        2,5
1129   Kuprianoff, J.
        Bound water in foods.
        In: Fundamental aspects of the dehydration of foodstuffs.
        pp 14-23.
        Conference in Aberdeen march 1958
        Society of Chemical Industry, London; 1958
        2
1130   Kuprianoff, J.
        Some factors influencing the reversibility of freeze-drying of
        foodstuffs.
        In: Fisher, F.R. (Ed.). Freeze Drying of Foods. pp 16-24.
        National Academy of Sciences-National Research Council,
        Washington D.C.; 1962
        2,5
1131   Kushner, D.J.
        Influence of solutes and ions on microorganisms.
        In: Hugo,W.B. (Ed.). Inhibition and Destruction of the
        Microbial Cell. pp 259-283.
        Academic Press, London; 1971
        4
1132   Kusik, C.L.; Meissner, H.P.
        Vapor pressures of water over aqueous solutions of strong
        electrolytes.
        Industrial and Engineering Chemistry. Product Research and
        Development 12 (1) 112-115; 1973
        3
1133   Kustova, L.V.; Gorshkov, V.I.
        Sorption of water vapor by ion exchangers studied by an
        isopiestic method.(Orig.russ.)
        Vestnik Moskovskogo Universiteta Khim Moscow 24 (4) 110-112;
        1969
        3,5
1134   Kvaale, O.; Dalhoff, E.
        Determination of the equilibrium relative humidity of foods.
        Food Technology 17 (5) 151-153; 1963
        3
1135   Labots, H.
        aw- und pH-Wert-Konzept für die Einteilung von
        Fleischerzeugnissen in verderbliche und lagerfähige Produkte.
        Die Fleischwirtschaft 61 (10) 1510-1517; 1981
        4
1136   Labuza, T.P.
        Sorption phenomena in foods.
        Food Technology 22 (3) 263-272; 1968
        1,2
1137   Labuza, T.P.
        Properties of water as related to the keeping quality of foods.
        In: Stewart, H.F. (Ed.). Proc. SOS/70. Third International
        Congress Food Science and Technol. pp 618-635. Chicago: Inst. of
        Food Technologists; 1971
        1,4

1138   Labuza, T.P.
       Kinetics of lipid oxidation in foods.
       CRC Critical Reviews in Food Technology 2 (3) 355-405; 1971
       4
1139   Labuza, T.P.
       Mechanisms of deterioration of intermediate moisture food
       systems.
       Nasa CR-114861, Washington; 1972
       4
1140   Labuza, T.P.
       Interpretation of sorption data in relation to the state of
       constituent water.
       In: Duckworth, R.B. (Ed.).
       Water Relations of Foods. pp 155-172.
       Academic Press, London; 1975
       1
1141   Labuza, T.P.
       Oxidative changes in foods at low and intermediate moisture
       levels.
       In: Duckworth, R.B. (Ed.).
       Water Relations of Foods. pp 455-474.
       Academic Press, London; 1975
       4
1142   Labuza, T.P.
       Sorption phenomena in foods: Theoretical and practical
       aspects.
       In: Rha, C. (Ed.). Theory Determination and Control of Physical
       Properties of Food Materials. pp 197-219.
       Reidel Press, Dordrecht; 1975
       1
1143   Labuza, T.P.
       The properties of water in relationship to water binding in
       foods. A review.
       Journal of Food Processing and Preservation 1 (2) 167-190;
       1977
       1,2
1144   Labuza, T.P.
       The effect of water activity on reaction kinetics of food
       deterioration.
       Food Technology 34 (4) 36-41, 59; 1980
       1,4
1145   Labuza, T.P.
       Enthalpy entropy compensation in food reactions.
       Food Technology 34 (2) 67-77; 1980
       2
1146   Labuza, T.P.
       Moisture gain and loss in packaged foods.
       Food Technology 36 (4) 92-97; 1982
       1,4
1147   Labuza, T.P.; Acott, K.; Tatini, S.R.; Lee, R.Y.; Flink, J.;
       McCall, W.J.
       Water activity determination: a collaborative study of
       different methods.
       Journal of Food Science 41 (4) 910-917; 1976
       3
1148   Labuza, T.P.; Busk, G.C.;
       An analysis of the water binding in gels.
       Journal of Food Science 44 (-) 1379-1385; 1979
       2
1149   Labuza, T.P.; Cassil, S.; Sinskey, A.J.
       Stability of intermediate moisture foods. II. Microbiology.
       Journal of Food Science 37 (1) 160-162; 1972
       4

1150 Labuza, T.P.; Chou, H.E.
Decrease of linoleate oxidation rate due to water at
intermediate water activity.
Journal of Food Science 39 (1) 112-113; 1974
4

1151 Labuza, T.P.; Heidelbaugh, N.; Silver, M.; Karel, M.
Oxidation at intermediate moisture contents.
Journal of the American Oil Chemists' Society 48 (2) 86-90; 1971
4

1152 Labuza, T.P.; Kamman, J.F.
Comparison of stability of thiamin salts at high temperature
and water activity.
Journal of Food Science 47 (2) 664-665; 1982
4

1153 Labuza, T.P. Kreisman, L.N.; Heinz, C.A.; Lewicki, P.P.
Evaluation of the Abbeon cup analyzer compared to the VPM and
Fett-Vos methods of water activity measurement.
Journal of Food Processing and Preservation 1 (1) 31-41; 1977
3

1154 Labuza, T.P.; Lewicki, P.P.
Measurement of gel water binding capacity by capillary suction
potential.
Journal of Food Science 43 (4) 1264-1269, 1273; 1978
2,3

1155 Labuza, T.P.; McNally, L.; Gallagher, D.; Hawkes, J.; Hurtado,F.
Stability of intermediate moisture foods. I. Lipid oxidation.
Journal of Food Science 37 (1) 154-159; 1972
4

1156 Labuza, T.P.; Mizrahi, S.; Karel, M.
Mathematical models for optimization of flexible film
packaging of foods for storage.
Transactions of the American Society of Agricultural Engineers
15 (-) 150-155; 1972
4,5

1157 Labuza, T.P.; Rutman, M.
Effect of surface-active agents on the sorption isotherms of a
model food system.
Canadian Journal of Chemical Engineering 46 (10) 364-368; 1968
1,2,5

1158 Labuza, T.P.; Saltmarch, M.
The nonenzymatic browning reaction as affected by water in
foods.
In: Rockland, L.B.; Stewart, G.F. (Edts.).
Water Activity: Influences on Foodquality. pp 605-650.
Academic Press, New York; 1981
4

1159 Labuza, T.P.; Sloan, A.E.; Acott, K.; Warmbier, H.C.
Intermediate moisture foods: chemical and nutrient stability.
Proc. IV. International Congress Food Science and Technology
1 (-) 546-557; 1974
4

1160 Labuza, T.P., Tannenbaum, S.R.; Karel, M.
Water content and stability of low-moisture,
intermediate-moisture foods.
Food Technology 24 (5) 543-550; 1970
1,4

1161 Labuza, T.P.; Tsuyuki, H.; Karel, M.
Kinetics of linoleate oxidation in model systems.
Journal of the American Oil Chemists' Society 46 (8) 409-416;
1969
4,5

1162 Lacroix, C.; Castaigne, F.
Détermination d'une fonction reliant l'activité de l'eau aux
teneurs en sel, sucre, protéines végétales et glycéro
emulsions cuites à base de viande de type Frankfurters.
Lebensmittel-Wissenschaft und -Technologie 16 (3) 129-134;
1983
4

1163 Lacroix, C.; Castaigne, F.
Evolution microbiologique en fonction de l'aw au cours de
l'entreposage des sausisses de type Frankfurters emballées
sous vide.
Lebensmittel-Wissenschaft und -Technologie 16 (3) 135-141;
1983
4

1164 Lafuente, B.; Carbonell, J.V.; Pinaga, F.
Influence of drying conditions on quality of freeze dried
green beans.
Revista de Agroquimica y Tecnologia de Alimentos
8 (-) 371-378; 1968
5

1165 Lafuente, B.; Pinaga, F.
Humedades de equilibrio de productos liofilizados.
Revista de Agroquimica y Tecnologia de Alimentos
6 (1) 113-117; 1966
5

1166 Laing, B.M.; Schlueter, D.L.; Labuza, T.P.
Degradation kinetics of ascorbic acid at high temperature and
water activity.
Journal of Food Science 43 (5) 1440-1443; 1978
4

1167 Lajolo, F.M.; Tannenbaum, S.; Labuza, T.P.
Reaction at limited water concentration.
2. Chlorophyll degradation.
Journal of Food Science 36 (6) 850-853; 1971
4

1168 La Mer, V.K.
The calculation of thermodynamic quantities from hysteresis
data.
Journal of Colloid and Interface Science 23 (-) 297-301; 1967
2

1169 Lancaster, E.B.
Thermodynamics of moist grain.
Die Stärke 34 (9) 296-299; 1982
2,5

1170 Landrock, A.H.; Proctor, B.E.
Measuring humidity equilibria.
Modern Packaging 24 (2) 123-130, 180; 1951
3,5

1171 Landrock, A.; Proctor, B.;
A new graphical interpolation method for obtaining humidity
equilibria data, with special reference to its role of food
packaging studies.
Food Technology 5 (8) 332-337; 1951
3,5

1172 Lang, K.W.
Physical, chemical and microbiological characterization of
polymer and solute bound water.
Ph.D. Thesis, University of Illinois, Urbana, Illinois, USA;
1981
1,2,4

1173 Lang, K.W.; McCune, T.D.; Steinberg, M.P.
      A proximity equilibration cell for rapid determination of
      sorption isotherms.
      Journal of Food Science 46 (3) 936-938; 1981
      3
1174 Lang, K.W.; Steinberg, M.P.
      Calculation of moisture content of a formulated food system to
      any given water activity.
      Journal of Food Science 45 (5) 1228-1230; 1980
      2,4,5
1175 Lang, K.W.; Steinberg, M.P.
      Predicting water activity from 0.30 to 0.95 of a
      multicomponent food formulation.
      Journal of Food Science 46 (3) 670-672, 680; 1981
      2,5
1176 Lang, K.W.; Steinberg, M.P.
      Linearization of the water sorption isotherm for homogeneous
      ingredients over aw 0.30 - 0.95.
      Journal of Food Science 46 (5) 1450-1452; 1981
      2,5
1177 Lang, K.W.; Steinberg, M.P.
      Characterization of polymer and solute bound water by pulsed
      NMR.
      Journal of Food Science 48 (2) 517-520, 538; 1983
      2
1178 Langmuir, I.
      The adsorption of gases on plane surfaces of glass, mica and
      platinum.
      Journal of the American Chemical Society 40 (-) 1361-1403;
      1918
      2
1179 Larher, Y.
      Transitions du premier ordre en phase adsorbée.
      Journal de Chimie, Physique, Physicochimie, Biologie
      65 (5) 974-976; 1968
      2
1180 Larmour, R.K.; Sallans, H.R.; Craig, B.M.
      Hygroscopic equilibrium of sunflowerseed flaxseed and
      soybeans.
      Canadian Journal of Research 22 Sec. F. (1) 1-8; 1944
      3,5
1181 Latyshev, V.P.; Agafonychev, V.P.
      Method for the calculation of water vapour pressure above
      foodstuffs with a wide range of water contents.
      Kholodil'naya Tekhnika 12 (-) 35-37; 1978
      2
1182 Lea, C.H.; Hannan, R.S.
      Studies on the reaction between proteins and reducing sugars
      in the "dry" state.
      I. The effect of activity of water of pH and temperature on
      the primary reaction between casein and glucose.
      Biochimica et Biophysica Acta 3 (-) 313-325; 1949
      4,5
1183 Lechert, H.
      Wasser als Strukturparameter im Lebensmittel - Untersuchung
      mittels Kernresonanzspektroskopie.
      40. Diskussionstagung, Forschungskreis der Ernährungsindustrie
      e.V. Hannover. pp 121-147; 1981
      2,5

1184 Lechert, H.T.
      Water binding on starch: NMR-studies on native and
      gelatinized starch.
      In: Rockland, L.B.; Stewart, G.F. (Edts.).
      Water Activity: Influences on Foodquality. pp 223-245.
      Academic Press, New York; 1981
      2
1185 Lechert, H.; Maiwald, W.; Köthe, R.; Basler, W.D.
      NMR-study of water in some starches and vegetables.
      Journal of Food Processing and Preservation 3 (-) 275-299;
      1980
      2
1186 Lee, B.B.; Kraft, A.A.
      Microbiology of intermediate moisture pork.
      Journal of Food Science 42 (3) 735-737; 1977
      4
1187 Lee, F.A.
      The effects of "bound" and "available" water on enzymic
      processes in wheat flour doughs.
      Food Technology of Australia 22 (9) 516-520; 1970
      1,2,4,5
1188 Lee, K.H.; Choi, H.Y.
      Water activity and pigment degradation in dried laver stored a
      room temperature.(Orig.kor.)
      Bulletin of the Korean Fisheries Society 6 (1/2) 27-36; 1973
      4
1189 Lee, R.Y.
      Influences of environmental factors on growth and production
      of enterotoxins by Staphylococcus aureus.
      Ph.D. Thesis, University of Minnesota, Minneapolis, Minnesota,
      USA; 1978
      4
1190 Lee, S.H.; Labuza, T.P.
      Destruction of ascorbic acid as a function of water activity.
      Journal of Food Science 40 (-) 370-373; 1975
      4,5
1191 Lee, S.; Dekay, H.G.; Banker, G.S.
      Effect of water vapor pressure on moisture sorption and the
      stability of Aspirin and ascorbic acid in tablet matrices.
      Journal of Pharmaceutical Sciences 54 (8) 1153-1158; 1965
      2,4,5
1192 Leeder, J.D.; Watt, I.C.
      The role of amino groups in water absorption by keratin.
      Journal of Physical Chemistry 69 (10) 3280-3284; 1965
      2
1193 Leeder, J.D.; Watt, I.C.
      The stoichiometry of water sorption by proteins.
      Journal of Colloid Interface Science 48 (2) 339-344; 1974
      2,5
1194 Legault, R.R.; Hendel, C.E.; Talburt, W.F.; Pool, M.F.
      Browning of dehydrated sulphited vegetables during storage.
      Food Technology 5 (-) 417-427; 1951
      4
1195 Legault, R.R.; Makower, B.; Talburt, W.F.
      Apparatus for measurement of vapor pressure.
      Analytical Chemistry 20 (5) 428-430; 1948
      3,5
1196 Leistner, L.
      Die Bedeutung des pH-Wertes, des Redoxpotentials und der
      Wasseraktivität für die Praxis der Fleischwarenherstellung.
      Archiv für Lebensmittelhygiene 21 (6) 121-126; 1970
      4

1197 Leistner, L.
Einfluß der Wasseraktivität von Fleischwaren auf die
Vermehrungsfähigkeit und Resistenz von Mikroorganismen.
Archiv für Lebensmittelhygiene 21 (12) 264-267; 1970
4
1198 Leistner, L.; Herzog, H.; Linke, H.
Wasseraktivität verschiedener Muskeln von Rind und Schwein.
Die Fleischwirtschaft 51 (4) 578-579; 1971
5
1199 Leistner, L.; Herzog, H.; Wirth, F.
Untersuchungen über die Wasseraktivität (aw-Wert) von
Rohwurst.
Die Fleischwirtschaft 51 (2) 213-216; 1971
1
1200 Leistner, L.; Karan-Djurdjic, S.
Beeinflussung der Stabilität von Fleischkonserven durch
Steuerung der Wasseraktivität.
Die Fleischwirtschaft 50 (11) 1547-1549; 1970
4
1201 Leistner, L.; Rödel, W.
Die Wasseraktivität (aw-Wert) als Kriterium für die
Beurteilung von Rohwurst.
Die Fleischwirtschaft 54 (6) 1039-1040; 1974
1
1202 Leistner, L.; Rödel, W.
Die Wasseraktivität bei Fleisch und Fleischwaren.
Deutsche Zeitschrift für Lebensmitteltechnologie
26 (6) 169-176; 1975
1,5
1203 Leistner, L. Rödel, W.
The significance of water activity for micro-organisms in meats.
In: Duckworth, R. (Ed.).
Water Relations of Foods. pp 309-323.
Academic Press, London; 1975
1,4
1204 Leistner, L.; Rödel, W.
The stability of intermediate moisture foods with respect to
micro-organisms.
In: Davies, R.; Birch, G.G.; Parker, K.J. (Edts.).
Intermediate Moisture Foods. pp 120-137.
Applied Science Publ., London; 1976
4
1205 Leistner, L.; Rödel, W.
Inhibition of microorganisms in food by water activity.
In: Skinner, F.A.; Hugo, W.B. (Edts.). Inhibition and
Inactivation of Vegetative Microbes. pp 219-237.
Academic Press, London, New York, San Francisco; 1976
4
1206 Leistner, L.; Rödel, W.
Microbiology of intermediate moisture foods.
Proc. International Meeting on Food Microbiology and
Tehnology. Tabiano B., Parma, Italy; 1978
4
1207 Leistner, L.; Rödel, W.; Krispien, K.
Microbiology of meat and meat products in high - and
intermediate-moisture ranges.
In: Rockland, L.B.; Stewart, G.F. (Edts.).
Water Activity: Influences on Foodquality. pp 855-916.
Academic Press, New York; 1981
4

1208 Leistner, L.; Wirth, F.
Bedeutung und Messung der Wasseraktivität (aw-Wert) von
Fleisch und Fleischwaren.
Die Fleischwirtschaft 52 (10) 1335-1337; 1972
1,3,4
1209 Leistner, L.; Wirth, F.; Vukovic, I.
SSP (Shelf Stable Products) - Fleischerzeugnisse mit Zukunft.
Die Fleischwirtschaft 59 (9) 1313-1318; 1979
1,4
1210 Le Maguer, M.
Rétention des matières volatiles sur des substrats à humidité
variable.
Dissertation, Université Paris; 1972
1,3,4
1211 Lenges, J.; Hinnekens, H.; Jacqmain, D.; Loncin, M.; Sacre, J.
Influence de l'humidité relative d'équilibre sur les
hydrolyses des substances pectiques.
Industries Alimentaires et Agricoles 86 (12) 1543-1550; 1969
4,5
1212 Lenzen, B.
Neue Messumformer zum Messen der relativen Luftfeuchtigkeit.
Verfahrenstechnik 9 (12) 602-604; 1975
3
1213 Leopold, G.H.; Johnston, J.
The vapor pressure of saturated aqueous solutions of
certain salts.
Journal of the American Chemical Society 49 (-) 1974; 1927
3
1214 Lerici, C.R.
Intermediate moisture foods.(Orig.ital.)
Industrie Alimentari 18 (10) 695-702; 1979
1,4
1215 Lerici, C.R.; Piva, M.
Water activity and food stability.(Orig.ital.)
Industrie Alimentari 20 (11) 763-769; 1981
1,4
1216 Lerici, C.R.; Piva, M.; Pinnavia, G.
Properties of intermediate-moisture food. I. Water activity and
chemical composition.(Orig.ital.)
Rivista della Societa Italiana di Scienza dell' Alimentazione
10 (10) 287-294; 1981
2,4
1217 Lespinasse, B.
Nouvelle méthode volumetrique pour la détermination
expérimentale des isothermes d'adsorption.
Compte rendu de l'Académie des Sciences de Paris,
Ser. C 267 (7) 359-361; 1968
3
1218 Lespinasse, B.
Sur une nouvelle méthode volumetrique pour la détermination
expérimentale des isothermes d'adsorption.
Compte rendu de l'Académie des Sciences de Paris,
Ser. C 267 (-) 1268-1270; 1968
2
1219 Lespinasse, B.; Brousse, E.
Nouvelle méthode de calcul des isothermes d'adsorption par
volumetrie.
Compte rendu de l'Académie des Sciences de Paris,
Ser. C 267 (8) 553-555; 1968
3

1220 Leung, H.K.
Capacity and force of water binding by carbohydrates and proteins
as determined by nuclear magnetic resonance.
Ph.D. Thesis, University of Illinois, Urbana, Illinois; 1974
2
1221 Leung, H.K.; Magnuson, J.A.; Bruinsma, B.L.
Water binding of wheat flour doughs and breads as studied by
deuteron relaxation.
Journal of Food Science 48 (1) 95-99; 1983
2
1222 Leung, H.; Morris, H.A.; Sloan, A.E.; Labuza, T.P.
Development of an intermediate-moisture processed cheese food
product.
Food Technology 30 (7) 42-44; 1976
4
1223 Leung, H.K.; Steinberg, M.P.
Water binding of food constituents as determined by NMR,
freezing, sorption and dehydration.
Journal of Food Science 44 (4) 1212-1216, 1220; 1979
2,5
1224 Leung, H.K.; Steinberg, M.P.; Wei, L.S.; Nelson, A.I.
Water binding of macromolecules determined by pulsed NMR.
Journal of Food Science 41 (2) 297-300; 1976
2
1225 Levi, A.; Ramirez-Martinez, I.R.; Padua, H.
Influence of heat and sulphur dioxide treatments on some
quality characteristics of intermediate moisture banana.
Journal of Food Technology 15 (-) 557-566; 1980
4
1226 Levine, A.S.; Fagerson, I.S.
A simplified moisture equilibrium apparatus.
Journal of the Technical Association of the Pulp and Paper
Industry 37 (7) 299; 1954
3
1227 Lewicki, P.P.
Znaczenie izoterm adsorpcji wody produktów spozywczych w
technologii zywnosci (Die Bedeutung der
Wasserdampf-Adsorptions-Isothermen von Lebensmitteln in der
Lebensmitteltechnik).
Przemysl Spozywczy 27 (11) 503-507; 1973
1
1228 Lewicki, P.P.
Powtarzalnosc statyczno - aksykatorowej metody wyznaczania
izoterm sorpcjiwody produktow spozywczych. (Repeatability of the
method for determination of water sorption isotherms of food
products)
Przemysl Spozywczy 30 (4) 141-143; 1976
5
1229 Lewicki, P.P.
A method to calculate constants in the BET equation applicable
to the type III isotherms of the BET classification.
Acta Alimentaria Polonica 3 (1) 67-77; 1977
2,5
1230 Lewicki, P.P.
Some observations on COST 90.
In: Jowitt, R.; Escher, F.; Hallström, B.; Meffert, H.F.T.; Spieß,
W.E.L.; Vos, G. (Edts.).
Physical Properties of Foods. pp 384-390.
Applied Science Publ., London, New York; 1983
4,5

1231 Lewicki, P.P..; Brzozowski, J.
Badanie izoterm adsorpcji wody wybranych produktów
spozywczych. (Wasserdampf-Adsorptionsisothermen ausgesuchter
Produkte).
Przemysl Spozywczy 27 (1) 18-21; 1973
5

1232 Lewicki, P.P.; Busk, G.C.; Labuza, T.P.
Measurement of gel water-binding capacity of gelatin, potato
starch, and carrageenan gels by suction pressure.
Journal of Colloid and Interface Science 64 (3) 501-509; 1978
2,3

1233 Lewicki, P.P.; Busk, G.C.; Peterson, P.L.; Labuza, T.P.
Determination of factors controlling accurate measurement of
aw by the VPM-technique.
Journal of Food Science 43 (1) 244-246; 1978
3

1234 Lewicki, P.P.; Lenart, A.
Wplyw procesu technologicznego na wlasciwosci adsorpcyjne
marchwi i porow suszonich.
(Der Einfluß des technologischen Prozesses auf die
Adsorptionseigenschaften von Karotten und getrocknetem Lauch.)
Przemysl Spozywczy 29 (2) 73-75; 1975
5

1235 Lewicki, P.P.; Lenart, A.
Wplyw wstepnego odwadniania osmotycznego na wlasciwosci
adsorpcyjne jablek suszonych owiewowo.
(Effect of preliminary osmotic dehydration on adsorptive
properties of apples dried in air current.)
Przemysl Spozywczy 31 (10) 394-397; 1977
5

1236 Lewicki, P.P.; Lenart, A.; Placzek, A.; Skrzeszewski, S.
Kinetyka sorpcji pary wodnej prrzez wybrane produkty
spozywcze.
(Kinetics of water vapor sorption by selected food products.)
Przemysl Spozywczy 31 (11) 428-432; 1977
3,5

1237 Lewicki, P.P.; Lorek, L.
Isotermy sorpcjiwody wybrenych produktow spozywczych.
(Sorption isotherms of water in selected food products.)
Przemysl Fermentacyjny i Rolny 21 (12) 25-27; 1977
5

1238 Li, K.C.; Heaton, E.K.; Boggess, T.S. jr.; Shewfelt, A.L.
Improving fruit quality with intermediate moisture process.
Food Product Development 8 (4) 52, 54; 1974
4

1239 Liang, S.C.
On the calculation of surface area.
Journal of Physical and Colloid Chemistry 55 (-) 1410-1412;
1951
3

1240 Lilly, V.; Birch, M.; Garscadden, B.
The preservation of spent brewers' grains by the application
of intermediate moisture food technology.
Journal of the Science of Food and Agriculture
31 (-) 1059-1065; 1980
4

1241 Ling, G.N.
The physical state of water in living cell and model systems.
Annals of the New York Academy of Sciences 125 (-) 401-417;
1965
2,5

1242 Ling, G.N.
The physical state of water in biological systems.
Food Technology 22 (10) 52-56; 1968
2

1243 Ling, G.
Water structure at the water-polymer interface.
In: Jellinek, H.H.G. (Ed.).
Water Structure at the Water Polymer Interface.
Plenum Press, New York; 1972
1

1244 Linko, P.
Water activity in cereal chemistry.(Orig.fin.)
Suomen Kemiustilehti 44 A (1) 1-8; 1971
1,5

1245 Linko, P.; Pollari, T.; Harju, M.; Heikonen, M.
Water sorption properties and the effect of moisture on
structure of dried milk products.
Lebensmittel-Wissenschaft und -Technologie 15 (1) 26-30; 1982
5

1246 Liou, J.K.; Bruin, S.
A model for watermigration in porous hygroscopic materials
caused by temperature gradients.
Lebensmittel-Wissenschaft und-Technologie 14 (4) 206-212; 1981
2,4

1247 Litzenberger, F.W.
Über die Verdichtungseigenschaften einiger Gemüsearten und ihr
Trocknungsverhalten in verdichtetem Zustand.
Industrielle Obst- und Gemüseverwertung 51 (18) 541-548; 1966
5

1248 Livingston, H.K.
Cross sectional areas of molecules adsorbed on solid surfaces.
Journal of the American Chemical Society 66 (-) 569-573; 1944
2

1249 Lladser, M.; Pinaga, F.
Criodeshidratación de aguacates.
I. Estudio sobre el comportamiento eutéctico e higroscópico
del aguacate liofilizado y ensayo de almacenamiento acelerado
del mismo.
Revista de Agroquimica y Tecnologia de Alimentos 15 (-) 547;
1975
5

1250 Loesecke, H.W. v.
Drying and dehydration of foods.
Reinhold Publ. Corporation, New York; 1955
1,5

1251 Lötzsch, R.; Leistner, L.
Überleben von Trichinella spiralis in Rohwurst und Rohschinken
in Abhängigkeit von der Wasseraktivität (aw-Wert).
Die Fleischwirtschaft 59 (2) 231-233; 1979
4

1252 Lötzsch, R.; Rödel, W.
Untersuchungen über die Lebensfähigkeit von Trichinella
spiralis in Rohwürsten in Abhängigkeit von der
Wasseraktivität.
Die Fleischwirtschaft 54 (-) 1203; 1974
4

1253 Lötzsch, R.; Trapper, D.
Bildung von Aflatoxinen und Patulin in Abhängigkeit von der
Wasseraktivität.
Die Fleischwirtschaft 58 (12) 2001-2007; 1978
4

1254 Loncin, M.
Influence de l'activité de l'eau sur les réactions chimiques
et biochimiques.
Dechema Monografie 56 (976/992) 193-205; 1965
3,4,5

1255 Loncin, M.
Die Grundlagen der Verfahrenstechnik in der
Lebensmittelindustrie.
Sauerländer, Aarau - Frankfurt/M.; 1969
1,2,4,5

1256 Loncin, M.
Basic principles of moisture equilibria.
In: Goldblith, S.A; Rey, L.; Rothmayr, W.W. (Edts.).
Freeze Drying and Advanced Food Technology. pp 599-617.
Academic Press, New York, London; 1975
1

1257 Loncin, M.
Wasseraktivität und halbfeuchte Lebensmittel.
36. Diskussionstagung Forschungskreis der Ernährungsindustrie
e.V. Hannover. pp 5-18; 1977
1,4,5

1258 Loncin, M.
Wasseraktivität - Bedeutung und Begriffsdefinition.
In: Loncin, M. (Ed.).
Hochschulkurs: Ausgewählte Themen der
Lebensmittelverfahrenstechnik "Wasseraktivität".
Institut für Lebensmittelverfahrenstechnik, Universität
Karlsruhe; 1980
1,2

1259 Loncin, M.
Food preservation by combined methods.
In: Seminario sobre ensenanze e investigacion en ingenieria de
alimentos. 13.-17. Nov. 1980, La Plata, Argentina, pp 73-79.
Departamentos de Asuntos Cientificos, Organizacion de los
Estados Americanos, Washington DC.; 1983
1,4

1260 Loncin, M.; Bimbenet, J.J.; Lenges, J.
Influence of the activity of water on the spoilage of
foodstuffs.
Journal of Food Technology 13 (2) 131-142; 1968
1,4,5

1261 Loncin, M.; Jacqmain, D.; Tutundjian-Provost, A.M.; Lenges,
J.P.; Bimbenet, J.J.
Influence de l'eau sur les réactions de Maillard.
Comptes-Rendus de l'Académie de Sciences de Paris
260 (-) 3208-3211; 1965
4

1262 Loncin, M.; Lenges, J.; Bimbenet, J.J.; Jacqmain, D.
L'humidité relative d'équilibre, son importance et sa mésure
Industries Alimentaires et Agricoles 82 (9/10) 923-927; 1965
1,3,5

1263 Loncin, M.; Weisser, H.
Die Wasseraktivität und ihre Bedeutung in der
Lebensmittelverfahrenstechnik.
Chemie-Ingenieur-Technik 49 (4) 312-319; 1977
1,5

1264 Lopez, F.L.C.; Christensen, C.M.
Invasion of and damage to bean seed by storage fungi.
Plant Desease Reporter 46 (11) 785-789; 1962
2,4,5

1265 Lotter, L.P.; Leistner, L.
Minimal water activity for enterotoxin A production and growth
of Staphylococcus aureus.
Applied and Environmental Microbiology 36 (2) 377-380; 1978
4
1266 Lourenco, O.B.; Pucci, J.R.
Equilibrium moisture of some cereals.
Ann. Asoc.quim. brasil. 13 (-) 15-18; 1954
5
1267 Lowe, E.; Durkee, E.L.; Farkas, D.F.; Silverman, G.J.
An idea for precisely controlling the water activity in testing
chambers.
Journal of Food Science 39 (5) 1072-1073; 1974
3
1268 Lozano, J.E.; Urbicain, M.J.; Rotstein, E.
Thermal conductivity of apples as a function of moisture
content.
Journal of Food Science 44 (1) 198-199; 1979
4
1269 Lubieniecki-von Schelhorn, M.
Vermehrung und Absterben von Mikroorganismen in Abhängigkeit
vom Milieu unter besonderer Berücksichtigung kombinierter
technologischer Einflüsse.
I. Vermehrung und Absterben von Mikroorganismen in
Abhängigkeit vom pH-Wert des Substrats (in wasserhaltigen
Substraten, bei Temperaturen zwischen dem Gefrierpunkt und
der beginnenden thermischen Abtötung.)
Chemie Mikrobiologie Technologie der Lebensmittel
1 (2) 89-95; 1972
4
1270 Lubieniecki-von Schelhorn, M.
Vermehrung und Absterben von Mikroorganismen in Abhängigkeit
vom Milieu unter besonderer Berücksichtigung kombinierter
technologischer Einflüsse.
II. Über das Verhalten von Bakteriensporen bei thermischen
Behandlungen in Abhängigkeit von der
Gleichgewichtsfeuchtigkeit der sie umgebenden Atmosphäre.
Chemie Mikrobiologie Technologie der Lebensmittel
1 (3) 138-144; 1972
4
1271 Lubienicki-von Schelhorn, M.
Vermehrung und Absterben von Mikroorganismen in Abhängigkeit
vom Milieu unter besonderer Berücksichtigung kombinierter
technologischer Einflüsse.
III. Einfluß der relativen Luftfeuchte auf die Hitzeresistenz
veschiedener Typen von Schimmelpilzsporen.
Chemie Mikrobiologie Technologie der Lebensmittel
2 (1) 26-32; 1973
4
1272 Lubieniecki-von Schelhorn, M.
Influence of relative humidity conditions on the thermal
resistance of several kinds of spores of molds.
Acta Alimentaria 2 (2) 163-171; 1973
4
1273 Lubieniecki-von Schelhorn, M.; Heiss, R.
The influence of relative humidity on the thermal resistance
of mold spores.
In: Duckworth, R.B. (Ed.).
Water Relations of Foods. pp 339-346.
Academic Press, London; 1975
4

-118-

1274 Lubieniecki-von Schelhorn, M.; Purr, A.
Beobachtungen über den lipolytischen Verderb von
Waffelfüllungen bei Gleichgewichtsfeuchtigkeiten unterhalb der
Wachstumsgrenze für Schimmelpilze.
Süßwaren 12 (3) 108-110; 1968
        12 (4) 165-166; 1968
4,5

1275 Lucae, T.
Zerstäubungstrocknung von Frucht- und Gemüseprodukten.
Zeitschrift für Lebensmittel-Technologie und Verfahrenstechnik
29 (7) 253-257; 1978
5

1276 Luck, W.A.P.
Structures of water in aqueous systems.
In: Rockland, L.B.; Stewart, G.F. (Edts.).
Water Activity: Influences on Foodquality. pp 407-434.
Academic Press, New York; 1981
2,5

1277 Lück, E.
Lebensmittel von mittlerer Feuchtigkeit.
Ernährungswirtschaft, Lebensmitteltechnik 20 (5) 346-352; 1973
1,4

1278 Lück, E.
Lebensmittel von mittlerer Feuchtigkeit.
Zeitschrift für Lebensmittel-Untersuchung und -Forschung
153 (1) 42-52; 1973
1,4

1279 Lück, E.
Lebensmittel von mittlerer Feuchtigkeit (Thema eines
Symposiums).
Ernährungswirtschaft ,Lebensmitteltechnik 23 (1) 15-16; 1976
4

1280 Lück, E.
Technologie der Lebensmittel von mittlerer Feuchtigkeit.
Gordian 77 (12) 344-346; 1977
4

1281 Lück, E.
Grundlagen zur Herstellung von Lebensmitteln von mittlerer
Feuchtigkeit.
Lebensmittel-Technologie 11 (3) 18-23; 1978
4

1282 Lück, W.
Feuchtigkeit - Grundlagen, Messen, Regeln.
Oldenbourg Verlag, München, Wien; 1964
3

1283 Lück, W.
Messung der Materialfeuchte.
Archiv für technisches Messen Blatt V 1281 - F2, 45-48; 1972
3

1284 Lüscher, M.;Giovanoli, R.; Hirter, P.
Untersuchungen der Hydratation von Collagen,
Sorptionsmessungen und Röntgenweitwinkeldiffraktion an
Tropocollagen.
Chimia 27 (2) 112-116; 1973
2,5

1285 Lüscher, M.; Rüegg, M.
Untersuchungen an Protein-Wasser-Systemen.
In: Marti, E.; Oswald, H.R.; Wiedemann, H.G. (Edts.).
Angewandte chemische Thermodynamik und Thermoanalytik.
pp 216-226. Birkhäuser Verlag; 1979
2

1286 Lüscher, M.; Rüegg, M.; Schindler, P.
Effect of hydration upon the thermal stability of
tropocollagen and its dependence on the presence of neutral
salts.
Biopolymers 13 (-) 2489-2503; 1974
2,5

1287 Lüscher-Mattli, M.; Rüegg, M.
Thermodynamic functions of biopolymer hydration.
I. Their determination by vapor pressure studies, discussed in
analysis of the primary hydration process.
Biopolymers 21 (-) 403-418; 1982
2

1288 Lüscher-Mattli, M.; Rüegg, M.
Thermodynamic functions of biopolymer hydration.
II. Enthalpy-entropy compensation in hydrophilic hydration
processes.
Biopolymers 21 (-) 419-429; 1982
2

1289 Lukanoff, T.; Baudisch, J.
The cellulose-water vapor system.
Zellstoff und Papier 16 (3) 88-93; 1967
3,5

1290 Lumry, R.
Some recent ideas about the nature of the interactions between
proteins and liquid water.
Journal of Food Science 38 (-) 744-755; 1973
1,2

1291 Lupin, H.M.; Boeri, R.L.; Moschiar, S.M.
Water activity and salt content relationship in moist salted
fish products.
Journal of Food Technology 16 (1) 31-38; 1981
4,5

1292 Luther, H.; Weber, E.; Schuster, E.
Ein Beitrag zur Charakterisierung der Wasserbindungskapazität
von Proteinen.
Die Nahrung 27 (3) 265-271; 1983
2

1293 Macchia, D.J.; Bettelheim, F.A.
Sorption of water vapor by amylose "B".
Journal of Polymer Science 2 (-) 1101-1104; 1964
2,5

1294 Mackay, B.H.; Stearn, A.E.; Downes, J.G.
Kinetics of water-vapor sorption in wool.
I. An improved sorption vibroscope.
Journal of the Textile Institute 60 (9) 359-377; 1969
2,3

1295 Mackay, B.H.; Downes, J.G.
Kinetics of water-vapor sorption in wool.
II. Results obtained with an improved sorption vibroscope.
Journal of the Textile Institute 60 (9) 378-394; 1969
2,3,5

1296 MacKenzie, A.P.
The physico-chemical environment during the freezing and
thawing of biological materials.
In: Duckworth, R.B. (Ed.).
Water Relations of Foods. pp 477-503.
Academic Press, London; 1975
3,5

1297 MacKenzie, A.P.; Luyet, B.J.
Water sorption isotherms from freeze-dried muscle fibers.
- Dependence on initial freezing treatment.
Cryobiology 3 (4) 341-344; 1967
5

1298 MacKenzie, A.P.; Rasmussen, D.H.
Interaction in the water-polyvinylpyrrolidone system at low
temperatures.
In: Jellinek, H.H.G. (Ed.).
Water Structure at the Water-Polymer Interface. p. 146.
Plenum Press, New York, London; 1972
3,5

1299 Madorsky, S.L.
Tungsten helical spring balance.
Review of Scientific Instruments 21 (4) 393-394; 1950
3

1300 Magrini, R.C.; Chirife, J.; Parada, J.L.
A study of Staphylococcus aureus growth in model systems and
processed cheese.
Journal of Food Science 48 (3) 882-885; 1983
4

1301 Mahler, K.
Apparatur zur kontinuierlichen Aufzeichnung von
Sorptionsisothermen.
Arbeitssitzung des Fachausschusses "Trocknungstechnik"
Göttingen; 1961
3

1302 Mahler, K.
Bestimmung der Restfeuchte über Gleichgewichtsdampfdruck und
Sorptionsisotherme.
Chemie-Ingenieur-Technik 36 (-) 461-463; 1964
3

1303 Maier, H.G.
Sorptionswärmen von flüchtigen Stoffen an einigen
Lebensmittelbestandteilen.
Journal of Chromatography 45 (-) 57-62; 1969
2,5

1304 Makhlouf, J.; Castaigne, F.; Goulet, J.; Vovan, X.
Oeufs liquides industriels à humidité intermédiaire.
I. Méthode de fabrication.
Canadian Institute of Food Science and Technology, Journal
16 (2) 123-129; 1983
4

1305 Makhlouf, J.; Castaigne, F.; Goulet, J.
Oeufs liquides industriels à humidité intermédiaire.
II. Comportement microbiologique au cours de l'entreposage.
Canadian Institute of Food Sience and Technology, Journal
16 (2) 130-135; 1983
4

1306 Makhlouf, J.; Castaigne, F.; Simard, C.
Oeufs liquides industriels à humidité intermédiaire.
III. Comportement fonctionnel et performance technologique au
cours de l'entreposage.
Canadian Institute of Food Science and Technology Journal
16 (2) 136-140; 1983
4

1307 Makower, B.
Vapor pressure of water adsorbed on dehydrated eggs.
Industrial and Engineering Chemistry 37 (10) 1018-1022; 1945
5

1308 Makower, B.
Determination of water in some dehydrated foods.
Advances in Chemistry Series 3 (-) 37-54; 1950
1,3,5

1309 Makower, B.; Dehority, G.L.
Equilibrium moisture content of dehydrated vegetables.
Industrial and Engineering Chemistry 35 (2) 193-197; 1943
2,3,5

1310 Makower, B.; Dye, W.B.
Equilibrium moisture content and crystallization of amorphous
sucrose and glucose.
Journal of Agricultural and Food Chemistry 4 (1) 72-76; 1956
4,5

1311 Makower, B.; Myers, S.
A new method for the determination of moisture in dehydrated
vegetables.
Proc. of the Institute of Food Technology. pp 156-164;
1943
3

1312 Makower, B.; Myers Chastain, S.; Nielsen, E.
Moisture determination in dehydrated vegetables. Vacuum oven
method.
Industrial and Engineering Chemistry 38 (7) 725-731; 1946
3

1313 Mallett, H.D.; Kohnen, J.B.; Surles, T.
Determination of water activity in intermediate moisture
pet foods by solvent extraction.
Journal of Food Science 39 (4) 847-848; 1974
3,4

1314 Maloney, J.F.; Labuza, T.P.; Wallace, D.H.; Karel, M.
Autoxidation of methyl linoleate in freeze-dried model
systems.
I. Effect of water on the autocatalyzed oxidation.
Journal of Food Science 31 (-) 878-883; 1966
4

1315 Mannheim, C.H.; Abraham, M.; Passy, N.
Water relations and stability of intermediate moisture foods.
Technion R.D. Foundation, Haifa; 1979
4,5

1316 Mannheim, C.H.; Passy, N.
Flow properties and water sorption of food powders.
I. Starches.
Lebensmittel-Wissenschaft und -Technologie 15 (4) 216-221;
1982
2,3,4,5

1317 Marchant, J.L.; Blanshard, J.M.V.
Changes in the birefringent characteristics of cereal starch
granules at different temperatures and water activities.
Die Stärke 32 (7) 223-226; 1980
4

1318 Marcos, A.; Alcala, M.; Leon, F.; Fernandez-Salguerro, J.;
Esteban, M.A.
Water activity and chemical composition of cheese.
Journal of Dairy Science 64 (4) 622-626; 1981
4,5

1319 Marcos, A.; Beltran de Heredia, F.H.; Alcala, M.;
Fernandez-Salguerro, J.; Esteban, M.A.; Leon, F.; Sanz, B.
Actividad de agua del yogur comercial.
Alimentaria 19 (134) 29-36; 1982
5

1320 Marcos, A.; Esteban, M.A.
Nomograph for predicting water activity of soft cheese.
Journal of Dairy Science 65 (9) 1795-1797; 1982
5

1321 Marcos, A.; Esteban, M.A.; Alcala, M.; Beltran de Heredia,
F.H.
Actividad del agua, pH y principales minerales del queso de
mahon.
Archivos de zootecnica 32 (122) 17-31; 1983
5

1322 Marcos, A.; Esteban, M.A.; Alcala, M.; Millan, R.
Prediction of water activity of San Simon cheese.
Journal of Dairy Science 66 (4) 909-911; 1983
5

1323 Marcos, A.; Esteban, M.A.; Fernandez-Salguero, J.
Actividad del agua y pH del queso manchego.
(Water activity and pH of the "Manchego" cheese)
Anal. Bromatol. 31 (1) 91-97; 1979
5

1324 Margaritis, A.; King, C.J.
Measurement of rates of moisture transport within the solid
matrix of hygroscopic porous materials.
Industrial Engineering Chemistry Fundamentals 10 (3) 510-515;
1971
2,3

1325 Margaritis, A.; King, C.J.
Factors governing terminal rates of freeze drying of poultry
meat.
Chemical Engineering Progress Symposium Series
67 (108) 112-121; 1971
2

1326 Marshall, B.J.; Murrell, W.G.; Scott, W.J.
The effect of water activity, solutes and temperature on the
viability and heat resistance of freeze-dried bacterial
spores.
Journal of General Microbiology 31 (-) 451-460; 1963
4

1327 Marshall, B.J.; Ohye, D.F.; Christian, J.H.B.
Tolerance of bacteria to high concentrations of NaCl and
glycerol in the growth medium.
Applied Microbiology 21 (2) 363-364; 1970
4

1328 Martin, E.C.
Some aspects of hygroscopic properties and fermentation of
honey.
Bee World 39 (7) 165; 1958
5

1329 Martin, S.
The control of conditioning atmospheres by saturated salt
solutions.
In: Wexler, A.; Wildhack, W.A. (Edts.).
Humidity and Moisture. Vol. III, pp 503-506. Reinhold Publ.
Corporation, New York; 1965
3

1330 Martinez, F.; Labuza, T.P.
Rate of deterioration of freeze-dried salmon as a function of
relative humidity.
Journal of Food Science 33 (-) 241-247; 1968
4,5

1331 Mashkovcev, M.; Volgunov, G.; Pokhno, M.
Influence des substances hygroscopiques accompagnant les
enzymes, sur les substances en équilibre avec l'humidité
relative.(Orig. russ.)
Biokhimija SSSR 16 (1) 24-28; 1951
4

1332 Masuzawa, M.; Sterling, C.
Gel water relationships in hydrophilic polymers:
Thermodynamics of sorption of water vapor.
Journal of Applied Polymer Science 12 (-) 2023-2032; 1968
2,3,5

1333 Matheson, N.A.
Absorption of atmospheric moisture by freeze dried pork and
fish.
Nature 184 (4703) 1949-1950; 1959
2
1334 Mathlouthi, M.; Conry, M.; Jaillant, G.; Maitenaz, P.C.
Water vapor sorption of Gruyere cheese.
Lebensmittel-Wissenschaft und -Technologie 13 (5) 264-268;
1980
3,5
1335 Mathlouthi, M.; Michel, J.F.; Maitenaz, P.C.
Study of some factors affecting water vapor sorption of
Gruyere cheese.
I. Proteolysis.
Lebensmittel-Wissenschaft und -Technologie 14 (3) 163-165;
1981
5
1336 Mattern, P.J.; Bishop, J.B.
A cabinet with controlled humidity to bring cereal samples to
constant moisture.
Cereal Science Today 18 (1) 8-10; 1973
3,5
1337 Matz, S.A.
Water in foods.
AVI Publ. Comp. Inc., Westport, Conn.; 1965
2,3,4
1338 Mauch, W.; Asseily, S.
Sorptionsverhalten von Fructose-, Glucose- und
Saccharoseschmelzen unterschiedlicher Erhitzungsgrade.
Forschungsbericht No.1, Institut für Zuckerindustrie Berlin;
1975
5
1339 Mayer, O.
Einfluß der Wasseraktivität auf den Ablauf chemischer
Reaktionen.
In: Loncin, M. (Ed.).
Hochschulkurs, Ausgewählte Themen der Lebensmitteltechnik
"Wasseraktivität".
Institut für Lebensmittelverfahrenstechnik Universität
Karlsruhe; 1980
4
1340 Mazza, G.
Water vapor equilibrium relationships of potato slices.
American Potato Journal 57 (3) 91-100; 1980
2,3,5
1341 Mazza, G.
Thermodynamic considerations of water vapor sorption by
horseradish roots.
Lebensmittel-Wissenschaft und -Technologie 13 (1) 13-17; 1980
2,5
1342 Mazza, G.
Moisture sorption isotherms of potato slices.
Journal of Food Technology 17 (1) 47-54; 1982
5
1343 Mazza, G.; LeMaguer, M.
Water sorption properties of yellow globe onion
(Allium cepa L.).
Canadian Institute of Food Science and Technology Journal
11 (4) 189-193; 1978
2,5

1344 Mazza, C.; LeMaguer, M.
      Flavor retention during dehydration of onion.
      2nd International Congress on Engineering and Food,
      August 21-27,Helsinki; 1979
      5
1345 McBain, J.W.
      An explanation of hysteresis in the hydration and dehydration
      of gels.
      Journal of the American Chemical Society 57 (-) 699-700; 1935
      2
1346 McBain, J.W.; Bakr, A.M.
      A new sorption balance.
      Journal of the American Chemical Society 48 (-) 690-695; 1926
      3
1347 McBain, J.W.; Good, S.J.; Bakr, A.M; Davies, D.P.; Willavoys,
      H.J.; Buckingham, R.
      The sorption of vapours by nitrocotton.
      Transaction of the Faraday Society 29 part 2, 1086-1100; 1933
      3,5
1348 McBean, D. McG.; Wallace, J.J.
      Stability of moist-pack apricots in storage.
      CSIRO Food Preservation Quarterly 27 (2) 29-35; 1967
      4,5
1349 McConnell, A.A.; Eastwood, M.A.; Mitchell, W.D.
      Physical characteristics of vegetable foodstuffs that could
      influence Bowel Function.
      Journal of the Science of Food and Agriculture
      25 (-) 1457-1464; 1974
      2
1350 McCune, T.D.; Lang, K.W.; Steinberg, P.
      Water activity determination with the proximity equilibration
      cell.
      Journal of Food Science 46 (6) 1978-1979; 1981
      3
1351 McCurdy, A.R.; Leung, H.K.; Swanson, B.G.
      Moisture equilibration and measurement in dry pinto beans
      (Phaseolus vulgaris).
      Journal of Food Science 45 (3) 506-508; 1980
      5
1352 McFarlane, J.A.
      Jamaica Marketing Departement Storage and Infestitation
      Division. p. 28; 1960
      cited in:
      Gough, M.C.; Lippiatt, G.A.
      Moisture humidity equilibria of tropical stored produce.
      Part III - Legumes, spices and beverages.
      Tropical Stored Products Information (35) 15-29; 1978
      5
1353 McLaren, A.D.; Rowen, J.W.
      Sorption of water vapor by proteins and polymers: a review.
      Journal of Polymer Science 7 (2/3) 289-324; 1952
      1,2,3,5
1354 McWeeny, D.J.
      Long term storage of some dry foods: a discussion of some of
      the principles involved.
      Journal of Food Technology 15 (-) 195-205; 1980
      1,4

1355 Measures, J.C.; Gould, G.W.
Interactions of micro-organisms with the environment of
intermediate moisture foods.
In: Davies, R.; Birch, G.G.; Parker, K.J. (Edts.).
Intermediate Moisture Foods. pp 281-297.
Applied Science Publ., London; 1976
4

1356 Meddings, P.J.; Potter, O.E.
The absorption of water by ground barley malt.
Cereal Chemistry 51 (1) 1-16; 1974
5

1357 Medema, J.; Houtman, J.P.W.
Brunauer-Emmett-Teller specific measurement of solids using
krypton.
Analytical chemistry 41 (1) 209-211; 1969
3

1358 Mellon, E.F.; Hoover, S.R.
Hygroscopicity of amino acids and its relationship to the
vapor phase water absorption of proteins.
Journal of the American Chemical Society 73 (-) 3879-3882;
1951
2,5

1359 Mellon, E.F.; Korn, A.H.; Hoover, S.R.
Water absorption of proteins.
I. Effect of free amino groups in casein.
Journal of the American Chemical Society 69 (4) 827-831; 1947
5

1360 Mellon, E.F.; Korn, A.H.; Hoover, S.R.
Water absorption of proteins.
II. Lack of dependence of hysteresis in casein on free amino
groups.
Journal of the American Chemical Society 70 (-) 1144-1146;
1948
2

1361 Mellon, E.F.; Korn, A.H.; Hoover, S.R.
Water absorption of proteins.
III. Contribution of the peptide group.
Journal of the American Chemical Society 70 (-) 3040-3044;
1948
5

1362 Mellon, E.F.; Korn, A.H.; Hoover, S.R.
Water absorption of proteins.
IV. Effect of physical structure.
Journal of the American Chemical Society 71 (-) 2761-2764;
1949
5

1363 Mellon, E.F.; Korn, A.H.; Kokes, E.L.; Hoover, S.R.
Water absorption of proteins.
VI. Effect of guanidino groups in casein.
Journal of the American Chemical Society 73 (-) 1870-1871;
1951
5

1364 Melpar, Inc.
Interrelationships between storage stability and moisture
sorption properties of dehydrated foods.
Contract DA 19-129-AMC-252 Final Report; 1965
4,5

1365 Menger, A.
Feuchtigkeitsprobleme bei der Lagerung von Dauerbackwaren.
Brot und Gebäck 8 (10) 156-159; 1954
1,5

1366 Menting, L.C.; Hoogstad, B.; Thijssen, H.A.C.
Diffusion coefficients of water and organic volatiles in
carbohydrate-water systems.
Journal of Food Technology 5 (2) 111-126; 1970
2,5

1367 Meredith, P.
Water absorption in wheat flour.
The Bakers Digest 43 (4) 42-45; 1969
2

1368 Middlehurst, J.; Parker, N.S.
Vermeidung von Kondenswasserbildung auf Konserven in
ISO-Behältern.
Die industrielle Obst- und Gemüseverwertung 63 (21) 585-592;
1978
4,5

1369 Migchelsen, C.; Berendsen, H.J.C.
Proton exchange and molecular orientation of water in hydrated
collagen fibers. A NMR study of H2O and D2O.
Journal of Chemical Physics 59 (1) 296-305; 1973
2

1370 Milacek, M.; Bohac, V.
Messen der Wasseraktivität von Schmelzkäse mittels einer
kryoskopischen Methode. (Orig. tschech.).
Protravinarska a Chladici Technika 10 (5) 158-160; 1980
3

1371 Militzer, K.E.; Siegel, B.
Zur Darstellung von Sorptionsisothermen.
Lebensmittel-Industrie 24 (2) 60-62; 1977
1,2,5

1372 Miller, B.S.; Kaslow, H.D.
Determination of moisture by nuclear magnetic resonance and
oven methods in wheat, flour, doughs and dried fruits.
Food Technology 17 (5) 650-653; 1963
3

1373 Miller, B.S.; Lee, M.S.; Hughes, J.W.; Pomeranz, Y.
Measuring high moisture content of cereal grains by pulsed
nuclear magnetic resonance.
Cereal Chemistry 57 (2) 126-129; 1980
3

1374 Miller, K.F.; Wright, P.G.
The equilibrium relative humidity of raw sugar.
Proc. Queensland Society of Sugar Cane Technologists.
pp 83-88; 1971
3,5

1375 Milner, M.; Geddes, W.F.
Grain storage studies.
III. The relation between moisture content, mold growth, and
respiration of soybeans.
Cereal Chemistry 23 (-) 225-247; 1946
4

1376 Miracco, J.L.; Alzamora, S.M.; Chirife, J.; Ferro-Fontan, C.
On the water activity of lactose solutions.
Journal of Food Science 46 (5) 1612-1613; 1981
3

1377 Mirna, A.
Versuche zur Berechnung der Wasseraktivität von Fleischwaren
aus analytischen Werten.
Die Fleischwirtschaft 50 (6) 831-833; 1970
1

1378 Mislivec, P.B.; Dieter, C.T.; Bruce, V.R.
Effect of temperature and relative humidity on spore
germination of mycotoxic species of Aspergillus and
Penicillium.
Mycologia 67 (-) 1187; 1975
4

1379 Misra, N.
Influence of temperature and relative humidity on fungal flora
of some spices in storage.
Zeitschrift für Lebensmittel-Untersuchung und -Forschung
172 (1) 30-31; 1981
4

1380 Miyata, A.; Watari, H.
A hygrometer which utilizes an anodic oxide film of aluminium.
In: Wexler, A.; Ruskin, R.E. (Edts.).
Humidity and Moisture. Measurement and Control in Science and
Industry. Vol. 1, pp 391-404. Reinhold Publ. Corporation,
New York; 1965
3,5

1381 Mizrahi, S.; Karel, M.
Moisture transfer in a packaged product in isothermal storage:
extrapolating data to any package-humidity combination and
evaluating water sorption isotherms.
Journal of Food Processing and Preservation 1 (3) 225-234;
1977
2,5

1382 Mizrahi, S.; Karel, M.
Accelerated stability tests of moisture sensitive products in
permeable packages by programming rate of moisture content
increase.
Journal of Food Science 42 (-) 958-963; 1977
2,4

1383 Mizrahi, S.; Karel, M.
Evaluation of kinetic model for reactions in moisture
sensitive products using dynamic storage conditions.
Journal of Food Science 43 (-) 750-753; 1978
4,5

1384 Mizrahi, S.; Kopelman, I.J.
Interrelationship of drained weight and water activity in
canned fruit systems.
Confructa 20 (1) 5-9; 1975
2,5

1385 Mizrahi, S.; Labuza, T.P.; Karel, M.
Computeraided predictions of extent of browning in dehydrated
cabbage.
Journal of Food Science 35 (-) 799-803; 1970
4,5

1386 Money, R.W.; Born, R.
Equilibrium humidity of sugar solutions.
Journal of the Science of Food and Agriculture 2 (4) 180-185;
1951
5

1387 Moor, H. de; Hendrickx, H.
Determination of the moisture content of dried milk and whey
powder with four methods.
International Dairy Congress (18th, Sydney) 1E: 436; 1970
3

1388   Morey, L.; Kilmer, H.; Selman, R.W.
        Relationship between moisture content of flour and humidity of
        air.
        Cereal Chemistry 24 (-) 364-371; 1947
        2,5
1389   Moreyra, R.; Peleg, M.
        Effect of equilibrium water activity on the bulk properties of
        selected food powders.
        Journal of Food Science 46 (6) 1918-1922; 1981
        4
1390   Moreyra-Sandoval, R.F.
        Mechanical behaviour of food powders under small compressive
        load.
        Ph.D. Thesis, University of Massachusetts, Amherst,
        Massachusetts; 1981
        4
1391   Morioka, Y.; Kobayashi, J.
        Sorption hysteresis and network structure of pores in porous
        substances (part 1) - Simulation of capillary condensation in
        adsorption process.
        Journal of the Chemical Society, Japan, Industrial Chemistry
        Sect. pp 157-164; 1979
        2
1392   Morris, H.J.; Wood, E.R.
        Influence of moisture content on keeping quality of dry beans
        Food Technology 10 (5) 225-229; 1956
        4
1393   Morrison, J.L.; Hanlan, J.F.
        The thermodynamic properties of the system wool keratin-water
        vapour.
        Proc. Intern. Congr. on Surface Activity Part 2.
        Solid-Gas Interface. pp 322-329; 1957
        2
1394   Morsi, M.K.S.; Sterling, C.; Volman, D.H.
        Sorption of water vapor by B pattern starch.
        Journal of Applied Polymer Science 11 (-) 1217-1225; 1967
        2,5
1395   Mossel, D.A.A.
        Occurence, prevention and monitoring of microbial quality los
        of foods and dairy products.
        Critical Reviews in Environmental Control 5 (-) 1-139; 1975
        4,5
1396   Mossel, D.A.A.
        Water and micro-organisms in foods. - A synthesis.
        In: Duckworth, R.B. (Ed.).
        Water Relations of Foods. pp 347-361.
        Academic Press, London; 1975
        4
1397   Mossel, D.A.A.
        Microbiological specifications for intermediate moisture food
        with special reference to methodology used for the assessment
        of compliance.
        In: Davies, R.; Birch, G.G.; Parker, K.J. (Edts.).
        Intermediate Moisture Foods. pp 248-259.
        Applied Science Publ., London; 1976
        4
1398   Mossel, D.A.A.; Ingram, H.
        The physiology of the microbial spoilage of foods.
        Journal of Applied Bacteriology 18 (-) 232-268; 1955
        4

1399 Mossel, D.A.A.; Kuijk, H.J.L. van
A new and simple technique for the direct determination of the equilibrium relative humidity of foods.
Food Research 20 (-) 415-423; 1955
3,4

1400 Motoki, M.; Torres, J.A.; Karel, M.
Development and stability of intermediate moisture cheese analogs from isolated soy bean proteins.
Journal of Food Processing and Preservation 6 (1) 41-53; 1982
4

1401 Moussa, A.E.
Kombinierter Effekt von Wasseraktivität, Strahlendosis und Temperatur während der Bestrahlung auf die Überlebensrate von drei Salmonellen-Serotypen.
Dissertation, Universität Hohenheim (Stuttgart); 1977
4,5

1402 Moussa, A.E.; Diehl, J.F.
Kombinierter Effekt von Strahlendosis, Bestrahlungstemperatur und Wasseraktivität auf die Überlebensrate von drei Salmonella Serotypen.
Archiv für Lebensmittelhygiene 30 (5) 171-176; 1979
4,5

1403 Mousseri, J.; Steinberg, M.P.; Nelson, A.I.; Wei, L.S.
Bound water capacity of corn starch and its derivatives by NMR.
Journal of Food Science 39 (1) 114-116; 1974
2

1404 Moy, J.H.;Chan, K.C.; Dollar, A.M.
Bound water in fruit products by the freezing method.
Journal of Food Science 36 (-) 498-499; 1971
2

1405 Mozumber, B.K.G.; Caroselli, N.E.; Albert, L.S.
Influence of water activity, temperature and their interaction on germination of Verticillium albo-atrum conidia.
Plant Physiology 46 (-) 347-349; 1970
4

1406 Muckle, T.B.; Stirling, H.G.
Review of the drying of cereals and legumes in the tropics.
Tropical Stored Product Information 22 (-) 11-30; 1971
1,2,5

1407 Müller, F.
Physikalisch-chemische Deutung der Mazerationsisotherme.
1. Mitt.: Vergleich der Mazerationsisotherme mit der Adsorptionsisotherme nach Freundlich.
Pharmazeutische Zentralhalle 107 (4) 270-274; 1968
2

1408 Müller, F.H.; Hellmuth, E.
Permeation, Diffusion, Adsorption und Quellung an Hochpolymeren.
I. Elektrisch registrierende Apparatur und einige Test-Messungen.
Kolloid-Zeitschrift 144 (1-3) 125-148; 1955
2,3

1409 Muffett, D.J.; Snyder, H.E.
Measurement of unfrozen and free water in soy proteins by differential scanning calorimetry.
Journal of Agricultural and Food Chemistry 28 (6) 1303-1305; 1980
2

1410 Multon, J.L.
Méthodes d'étude des isothermes de sorption de vapeur d'eau et
mésure de l'humidité relative d'équilibre d'un produit.
Proc. Intermediate moisture foods. CPCIA - Seminar 5
Paris; 1975
1,3

1411 Multon, J.L.
The state of water in less hydrated biological and natural
products (water corresponding to that extracted during
secondary drying and to residual moisture, in lyophilized
products).
7eme Cours Internationaux de Lyophilisation, INSAL, Lyon; 1977
2

1412 Multon, J.L.
Normalisation et Recherche: le dosage de l'eau dans les
céréales et son complément indispensable, le circuit d'analyse.
Courrier de la Normalisation 268, XI-XII, 441-477; 1979
3

1413 Multon, J.L.
L'état actuel des travaux de la commission "aliments à
humidité intermédiaire" du C.N.E.R.N.A..
Industries Alimentaires et Agricoles 98 (4) 291-302; 1981
1,4

1414 Multon, J.L.
Les interactions entre l'eau et les constituants des grains,
graines et produits derivés.
In: Conservation et stockage des grains et graines.
Tec. et Doc. APRIA, Paris; 1982
1,2,4,5

1415 Multon, J.L.; Bizot, H.
Methods for studying water vapor sorption and water activity
of materials.
In: Walters, A.H. (Ed.).
Biodeterioration Investigation Techniques.
Applied Science Publ., London; 1976
1,3,5

1416 Multon, J.L.; Bizot, H.
Aliments à humidité intermédiaire et détermination de
l'activité de l'eau.
Annales de la Nutrition et de l'Alimentation 32 (2/3) 631-654;
1978
1,4,5

1417 Multon, J.L.; Bizot, H.; Doublier, J.L.; Lefebvre, J.; Abbott, I
Effect of water activity and sorption hysteresis on
rheological behaviour of wheat kernels.
In: Rockland, L.B.; Stewart, G.F. (Edts.).
Water Activity: Influences on Foodquality. pp 179-198.
Academic Press, New York; 1981
2,5

1418 Multon, J.L.; Bizot, H.: Martin, G.
Méthodes de mésure et de controle des états de l'eau dans les
aliments.
Techniques d'Analyse et de controle dans les Industries
Agro-Alimentaires; Vol. IV, A.P.R.I.A.; 1980
1,3

1419 Multon, J.L.; Gadet, M.H.; Martin, G.
Le dosage de l'eau dans les céréales: Methode de référence et
humidimètres.
Revue Métrologie Pratique et Légale (mai) 342-357; 1978
3

1420 Multon, J.L.; Guilbot, A.
Méthode de détermination rapide des isothermes de sorption de vapeur d'eau.
Annales de Technologie Agricole 16 (1) 5-25; 1967
3,5

1421 Multon, J.L.; Guilbot, A.
La régulation de l'humidité relative dans les cellules de petit volume.
Industries Alimentaires et Agricoles 85 (4) 405-413; 1968
2,3

1422 Multon, J.L.; Guilbot, A.
Water activity in relation to the thermal inactivation of enzymic proteins.
In: Duckworth, R. (Ed.).
Water Relations of Foods. pp 379-396.
Academic Press, London; 1975
4

1423 Multon, J.L.; Martin, G.
Le dosage de l'eau dans les grains; étude de la répétabilité de la méthode de référence pratique de dosage de l'eau dans les céréales: Calculs théoriques et évaluation expérimentale.
Revue de Métrologie Pratique et Légale (juin) 450-461; 1978
3

1424 Multon, J.L.; Martin, G.
International standardized methods and moisture meters for determining moisture content in cereal grains.
Cereal Food World 24 (11) 548-558; 1979
3

1425 Multon, J.L.; Savet, B.; Bizot, H.
A fast method for measuring the activity of water in foods.
Lebensmittel-Wissenschaft und -Technologie 13 (5) 271-273; 1980
1,3,5

1426 Multon, J.L.; Trentesaux, E.; Guilbot, A.
Méthode rapide de détermination des isothermes de sorption de l'eau par les solides.
Lebensmittel-Wissenschaft und -Technologie 4 (6) 184-189; 1971
3,5

1427 Murray, D.G.; Luft, L.R.
Low-D.E. corn starch hydrolysates.
Food Technology 27 (3) 32-40; 1973
5

1428 Murrell, W.G.; Scott, W.J.
The heat resistance of bacterial spores at various water activities.
Journal of General Microbiology 43 (-) 411; 1966
4

1429 Naesens, W.; Bresseleers, G.; Tobback, P.
A method for the determination of diffusion coefficients of food components in low- and intermediate moisture systems.
Journal of Food Science 46 (5) 1446-1449; 1981
2

1430 Naesens, W.; Bresseleers, G.; Tobback, P.
Diffusional behavior of tripalmitin in a freeze-dried model system at different water activities.
Journal of Food Science 47 (4) 1245-1249; 1982
2,5

1431 Nagashima, N.; Suzuki, E.
Pulsed NMR and state of water in foods.
In: Rockland, L.B.; Stewart, G.F. (Edts.).
Water Activity: Influences on Foodquality. pp 247-264.
Academic Press, New York; 1981
2

1432 Nagiev, M.F.; Ibragimov, C.S.; Gadzhieva, E.A.
      Sorption-structural isotherms.(Orig. russ.)
      Azerbajdzan Kimja Zurnaly 6 (-) 49-53; 1966
      2
1433 Nagy, N.
      Influence des hétérogenités sur l'équilibre de sorption.
      dans les systèmes gel polymère-mélange binaire.
      Magy-Kem. Foly 87 (-) 462-474; 1981
      2
1434 Nakajima, T.; Sugai, K.; Ito, H.
      Sorption of water by chitosan.(Orig. jap.)
      Kobunshi Robunsha 37 (10) 705-710; 1980
      5
1435 Nakano, H.: Yasui, T.
      Denaturation of myosin-ATP-ase as a function of water
      activity.
      Agricultural and Biological Chemistry 40 (1) 107-113; 1976
      4
1436 Nakano, H.; Yasui, T.
      Inactivation of myosin B and myofibrillar ATPases as a
      function of water activity and the inhibitory effect of sucros
      on the dehydration induced inactivation of myosin ATPase.
      (Orig. jap.)
      Nippon Chikusan Gakkai Ho 50 (3) 161-172; 1979
      4
1437 Nara, S.
      On the water sorption in starch granules. (Orig. jap.)
      Journal of the Japanese Society of Starch Science, Tokyo
      28 (1) 24-32; 1981
      2,5
1438 Nara, S.; Yabumoto, Y.; Yamaguchi, K.; Maeda, I.
      Sorption water of starch granules.(Orig. jap.)
      Journal of the Agricultural Chemical Society of Japan (Nihon
      Nogei Kagakkai-shi) 43 (8) 570-574; 1969
      5
1439 Nara, T.
      Study on the heat of wetting of starch.
      Starch/Stärke 31 (4) 105-108; 1979
      2
1440 Neale, R.J.; Obanu, Z.A.; Biggin, R.J.; Ledward, D.A.;
      Lawrie, R.A.
      Protein quality and iron availability of intermediate moisture
      meat stored at 37 °C.
      Annales de la Nutrition et de l'Alimentation 32 (-) 587-596;
      1978
      4
1441 Neale, S.M.; Stringfellow, W.A.
      The primary sorption of water by cotton.
      Transaction of the Faraday Society 37 (-) 525-532; 1941
      2,3,5
1442 Nekryach, E.F.; Samchenko, Z.A.
      Sorption isotherms of water vapor and wetting heats of
      poly(w-pelargonamide).(Orig. russ.)
      Fiz.-Khim. Mekh. Liofil'nost Dispersnykh Sist. 63-68; 1968
      2,5
1443 Nellist, M.E.; Hughes, M.
      Physical and biological processes in the drying of seed.
      Seed Science and Technology 1 (-) 613-643; 1973
      2,5

1444 Nelson, T.J.
Hygroscopicity of sugar and other factors affecting retention
of quality.
Food Technology 3 (-) 347-351; 1949
4,5

1445 Nelson, J.; Schubert, M.
Water concentration and the rate of hydrolysis of sucrose by
invertase.
Journal of the American Chemical Society 50 (-) 2188-2193;
1928
4

1446 Nemethy, G.; Scheraga, H.A.
The Structure of water and hydrophobic bonding in proteins.
I. A model for the thermodynamic properties of liquid water.
Journal of Chemical Physics 36 (12) 3382-3400; 1962
2

1447 Nemethy, G.; Scheraga, H.A.
The structure of water and hydrophobic bonding in proteins.
II. Model for the thermodynamic properties of aqueous
solutions of hydrocarbons.
Journal of Chemical Physics 36 (12) 3401-3417; 1962
2

1448 Nemethy, G.; Scheraga, H.
The structure of water and hydrophobic bonding in proteins.
III. The thermodynamic properties of hydrophobic bonds in
proteins.
The Journal of Physical Chemistry 66 (-) 1773-1789; 1962
2

1449 Nemitz, G.
Über die Wasserbindung durch Eiweißstoffe und deren Verhalten
während der Trocknung.
Dissertation, Technische Hochschule Karlsruhe; 1961
2,3,5

1450 Nemitz, G.
Physikalisch-chemische Grundlagen der Gefriertrocknung von
Proteinen.
Proc. 5. Gefriertrocknungstagung Köln.
Leybold Hochvakuum Anlagen GmbH, Köln-Bayental; 1962
2,5

1451 Nemitz, G.
Die hygroskopischen Eigenschaften von Stärken und
Stärkeprodukten.
Die Stärke 14 (8) 276-278; 1962
3,5

1452 Nemitz, G.
Hygroskopische Eigenschaften von Trockengemüsen.
Die industrielle Obst- und Gemüseverwertung 17 (6) 162-163;
1962
3,5

1453 Nemitz, G.
Die hygroskopischen Eigenschaften getrockneter Lebensmittel.
Zeitschrift für Lebensmittel-Untersuchung und -Forschung
123 (1) 1-5; 1963
3,5

1454 Neuber, E.E.
Evaluation of critical parameters for developing moisture
sorption isotherms of cereal grains.
In: Rockland, L.B.; Stewart, G.F. (Edts.).
Water Activity: Influences on Foodquality. pp 199-222.
Academic Press, New York; 1981
1,3,5

1455 Newns, A.C.
Sorption and desorption kinetics of the cellulose and water system.
I. Successive sorptions from dryness.
Transaction of the Faraday Society 64 (11) 3147-3151; 1968
2,5

1456 Ngoddy, P.O.
A generalized theory of sorption phenomena in biological materials.
Ph.D. Thesis, Agricultural Engineering Department, Michigan State University, East Lansing; 1969
1,2,3,4,5

1457 Ngoddy, P.O.; Bakker-Arkema, F.W.
A generalized theory of sorption phenomena in biological materials.
I. The isotherm equation.
Transactions of the American Society of Agricultural Engineers 13 (-) 612-617; 1970
2,5

1458 Ngoddy, P.O.; Bakker-Arkema, F.W.
A theory of sorption hysteresis in biological materials.
Journal of Agricultural Engineering Research 20 (-) 109-121; 1975
3,5

1459 Niedik, E.A.
Über Eigenschaften amorpher Zucker und ihre Bedeutung für die Lebensmittel- und Süßwarenindustrie.
39. Diskussionstagung, Forschungskreis der Ernährungsindustrie e.V. Hannover, pp 153-175; 1980
5

1460 Niedik, E.A.
Über Eigenschaftsunterschiede zwischen der kristallinen und amorphen Form von Saccharose und Lactose.
Zeitschrift für Lebensmittel-Technologie und -Verfahrenstechnik 33 (3) 173-185; 1982
2,5

1461 Niedik, E.A.; Babernics, L.
Aromasorptionseigenschaften von amorpher Saccharose und Lactose.
Gordian 79 (2) 35-44; 1979
2

1462 Niediek, E.A.; Barbernics, L.
Amorphisierung von Zucker durch das Feinwalzen von Schokoladenmassen.
Gordian 80 (11) 267-269; 1980
2

1463 Nikitina, L.M.
Thermodynamic parameters and coefficients of masstransfer of humid materials. (Orig. russ.)
Energie-Verlag, Moskau; 1968
2,5

1464 Nip, W.K.
Development and storage stability of drum-dried guava and papaya-taro flakes.
Journal of Food Science 44 (1) 222-225; 1979
4,5

1465 Nirkko, P.
Wasserdampfsorption von Kartoffelstärke, ihren Bestandteilen und Abbauprodukten.
Inaugural-Dissertation, Universität Bern; 1973
2,3,5

1466 Nirkko, P.; Gál, S.; Giovanoli, R.; Signer, R.
      Der Einfluß des Polymerisationsgrades von Oligosacchariden auf
      die Wasserdampfsorption und ihre Hysterese.
      Die Stärke 27 (8) 278-280; 1975
      5
1467 N.N.
      Relative humidity over salt solutions.
      Tappi Data Sheet 109, Technical Association of the Pulp and
      Paper Industry, New York; 1944
      3
1468 N.N.
      Automatic dewpoint hygrometer.
      Engineering 173 (5) 634-635; 1952
      3
1469 N.N.
      Relative humidity of air over saturated solutions of salts.
      British Standard 3718; 1964
      Specification for Laboratory Humidity Ovens (Non-Injection
      Type), p. 19; 1964
      3
1470 N.N.
      Intermediate-moisture foods retain fresh properties.
      Food Engineering 42 (5) 119-120; 1970
      4
1471 N.N.
      Recommended practice for maintaining constant relative
      humidity by means of aqueous solutions.
      ASTM, Designation: E 104-51; 1971
      3
1472 N.N.
      Water activity - an important parameter in the food industry.
      Food Trade Review 49 (8) 451; 1979
      1
1473 N.N.
      Wasseraktivität; Grundlagen und Meßtechnik.
      Firmenprospekt Rotronic GmbH; 1981
      1
1474 N.N.
      Determination of water activity of raw sausages as a function
      of moisture and salt contents.(Orig. fr.)
      Viandes et Produits Carnes 2 (8) 11-13; 1981
      2,5
1475 Noguchi, H.
      Hydration around hydrophobic groups.
      In: Rockland, L.B.; Stewart, G.F. (Edts.).
      Water Activity: Influences on Foodquality. pp 281-293.
      Academic Press, New York; 1981
      2
1476 Norrish, R.S.
      An equation for the activity coefficients and equilibrium
      relative humidities of water in confectionary syrups.
      Journal of Food Technology 1 (1) 25-39; 1966
      3,5
1477 Norrish, R.S.
      Equilibrium relative humidity of confectionary syrups.
      In: Hoynak, P.; Bollenback, G. (Edts.).
      This is Liquid Sugar. pp 232-236.
      2nd edition, Yonkers, Refined Syrups and Sugars Inc.; 1966
      2,5

1478 Northcote, D.H.
      The sorption of water vapour by yeast cell wall and other
      polysaccharides.
      Biochimica et Biophysica Acta 11 (-) 471-479; 1953
      2,5

1479 Northolt, M.D.
      Effects of water activity on microorganisms in foods. I.
      Voedingsmiddelentechnologie 12 (19) 23-27; 1979
      4,5

1480 Northolt, M.D.; Egmond, H.P. van; Paulsch, W.E.
      Differences between Aspergillus flavus strains in growth and
      aflatoxin B1 production in relation to water activity and
      temperature.
      Journal of Food Protection 40 (11) 778-781; 1977
      4

1481 Northolt, M.D.; Egmond, H.P. van; Paulsch, W.E.
      Patulin production by some fungal species in relation to water
      activity and temperature.
      Journal of Food Protection 41 (-) 885-890; 1978
      4

1482 Northolt, M.D.; Egmond, H.P. van; Paulsch, W.E.
      Penicillic acid production by some fungal species in relation
      to water activity and temperature.
      Journal of Food Protection 42 (6) 476-484; 1979
      4

1483 Northolt, M.D.; Egmond, H.P. van; Paulsch, W.E.
      Ochratoxin A production by some fungal species in relation to
      water activity and temperature.
      Journal of Food Protection 42 (6) 485-490; 1979
      4

1484 Northolt, M.D.; Heuvelman, C.J.
      Effects of water activity on microorganisms in foods. II.
      Voedingsmiddelentechnologie 12 (20) 49-52; 1979
      4,5

1485 Northolt, M.D.; Heuvelman, C.J.
      De zoutkristalvervloeiingstest: een eenvoudige methode voor
      het controleren van de wateractiviteit van voedingsmiddelen.
      (Der Salzkristall-Auflösungstest: Eine einfache Methode zum
      Kontrollieren der Wasseraktivität von Lebensmitteln)
      Voedingsmiddeltechnologie 15 (1) 11-15; 1982
      3,5

1486 Northolt, M.D.; Heuvelman, C.J.
      The salt crystal liquefaction test. A simple method for
      testing the water activity of foods.
      Journal of Food Protection 45 (6) 537-540, 546; 1982
      3,5

1487 Northolt, M.D.; Verhulsdonk, C.A.H.; Soentoro, P.S.S.;
      Paulsch, W.E.
      Effect of water activity and temperature on aflatoxin
      production by Aspergillus parasiticus.
      Journal of Milk and Food Technology 39 (3) 170-174; 1976
      4

1488 Notter, G.K.; Taylor, D.H.; Brekke, J.E.
      Pineapple juice powder.
      Food Technology 12 (-) 363-366; 1958
      4,5

1489 Notter, G.K.; Taylor, D.H.; Downes, N.J.
      Orange juice powder. Factors affecting storage stability.
      Food Technology 13 (-) 113-118; 1959
      4,5

1490   Notter, G.K.; Taylor, D.H.; Walker, L.H.
       Stabilized lemonade powder.
       Food Technology 9 (-) 503-505; 1955
       5
1491   Novitskaya, V.E.; Ovrutskaya, I.Y.
       Determination of water activity in dried potato granules.
       (Orig. russ.)
       Konservnaya i Ovoshchesushil'naya Promyshlennost' (5) 35-36;
       1978
       3,5
1492   Novokshonov, Y.I.
       Operational prediction of the water-adsorption capacity of
       wheat flour.
       Izvestiya Vysshikh Uchebnykh Zavadenil, Pishchevaya Tekhnologiya
       3 (-) 101-103; 1981
       5
1493   Obanu, Z.A.; Ledward, D.A.; Lawrie, R.A.
       The proteins of intermediate moisture meat stored at tropical
       temperature.
       III. Differences between muscles.
       Journal of Food Technology 11 (-) 187-196; 1976
       4
1494   Obermiller, J.
       Die Einstellung von Luft auf bestimmte Trocknungs- oder
       Feuchtigkeitsgrade mit Hilfe von Salzen und ähnlichen Stoffen.
       Zeitschrift für Physikalische Chemie 109 (-) 145-164; 1924
       3
1495   Oberol, A.S.; Fuller O.M.; Kelly, J.F.
       Methods of interpreting transient response curves from dynamic
       sorption experiments.
       Industrial and Engineering Chemistry Fundamentals
       19 (1) 17-21; 1980
       2
1496   O'Brien, F.E.M.
       The control of humidity by saturated salt solutions.
       Journal of Scientific Instruments 25 (-) 73-76; 1948
       3
1497   Odamtten, G.T.; Langerak, D.I.
       Moisture sorption of two maize varieties kept under different
       ambient relative humidities.
       Report No. 9, International Facility for Food Irradiation
       Technology (IFFIT), Wageningen; 1980
       3,5
1498   Ogiwara, Y.; Kubota, H.; Hayashi, S.; Mitomo, H.
       Studies of water adsorbed on cellulosic materials by a
       high-resolution NMR spectrometer.
       Journal of Applied Polymer Science 13 (8) 1689-1696; 1969
       2
1499   Okamura, T.
       A study of the drying characteristics of soybeans in relation
       to minute structure.
       Research Bulletin of Obihiro Zootechnical University
       5 (5) 767-779; 1968
       5
1500   Okamura, T.
       Studies on the state of water in soybean seed in relation to
       moisture content.
       Research Bulletin of Obihiro Zootechnical University,
       Series I, 8 (2) 89-145; 1973
       2,5

1501   Okamura, T.; Steinberg, M.P.; Tojo, M.; Nelson, A.I.
       Water binding by soy flours as measured by wide line NMR.
       Journal of Food Science 43 (2) 553-555, 559; 1978
       2
1502   Okwelogu, T.N.
       Maximum safe moisture content for storing Teff (Eragrostis
       abyssinica S. and Sarawak Illipe nut (shorea gysbertsiana
       Burck).
       Journal of Stored Products Research 5 (2) 169-172; 1969
       5
1503   Okwelogu, T.N.
       Guidance in the selection of moisture meters for durable
       agricultural produce.
       Tropical Stored Product Information (21) 19-29; 1971
       3
1504   Okwelogu, T.N.; MacKay, F.J.
       Cashewnut moisture relations.
       Journal of the Science of Food and Agriculture
       20 (12) 697-702; 1969
       5
1505   Olynyk, P.; Gordon, A.R.
       The vapor pressure of aqueous solutions of sodium chloride at
       20, 25 and 30 °C for concentrations from 2 molal to saturation.
       Journal of the American Chemical Society 65 (-) 224; 1943
       3
1506   Ooraikul, B.
       Gas packaging for a bakery product.
       Canadian Institute of Food Science and Technology, Journal
       15 (4) 313-315; 1982
       4,5
1507   Orr, C.
       Der derzeitige Stand der Methoden zur Bestimmung von Pulver
       Oberflächen.
       Chemie-Ingenieur-Technik 48 (8) 680-689; 1976
       1
1508   Orr, W.J.C.
       The adsorption of non-polar gases on alcali halide crystals.
       Proc. of the Royal Society of London A 173 (-) 349-367; 1939
       2,3
1509   Orth, R.
       Der Einfluß der Wasseraktivität auf die Sporenkeimung bei
       Aflatoxin, Sterigmatocystin und Patulin bildenden Schimmel-
       pilzarten.
       Lebensmittel-Wissenschaft und -Technologie 9 (3) 156-159; 1976
       4,5
1510   Oswin, C.R.
       The kinetics of package life. III. The isotherm.
       Journal of the Society of Chemical Industry
       65 (-) 419-421; 1946
       2,4
1511   Oswin, C.R.
       Food packaging in relation to moisture exchange.
       Chemistry and Industry 24 (-) 1042-1044; 1976
       2,4,5
1512   Othmer, D.F.
       Correlating vapor pressure and latent heat data. A new plot.
       Industrial and Engineering Chemistry 32 (6) 841-856; 1940
       2
1513   Othmer, D.F.; Huang, H.N.
       Correlating vapor pressure and latent heat data.
       Industrial and Engineering Chemistry 57 (10) 42-48; 1965
       2

1514   Othmer, D.F.; Sawyer, F.G.
        Correlating adsorption data. Temperature, pressure,
        concentration, heat.
        Industrial and Engineering Chemistry 35 (12) 1269-1276; 1943
        2
1515   Otsuka, A.; Wakimoto, T.; Takeda, A.
        Moisture sorption and volume expansion of lactose anhydrate
        tablets. (Orig. jap.)
        Yakugaku Zasshi 96 (3) 351-355; 1976
        2,4
1516   Otsuka, A.; Wakimoto, T.; Takeda, A.
        Moisture sorption and volume expansion of amorphous lactose
        tablets.
        Chemical and Pharmaceutical Bulletin 26 (3) 967-971; 1978
        2,4
1517   Oxley, T.A.
        The scientific principles of grain storage.
        Northern Publ. Co., Liverpool; 1948
        1,2,4
1518   Padival, R.A.; Ranganna, S.; Manjrekar, S.P.
        Stability of pectins during storage.
        Journal of Food Technology 16 (4) 367-378; 1981
        4,5
1519   Palmer, K.J.; Dye, W.B.; Black, D.
        X-ray diffractometer and microscopic investigation of
        crystallization of amorphous sucrose.
        Journal of Agricultural and Food Chemistry 4 (1) 77-81; 1956
        4
1520   Palmer, K.J.; Elsken, R.H.
        Determination of water by nuclear magnetic resonance in
        hygroscopic materials containing soluble solids.
        Journal of Agricultural and Food Chemistry 4 (2) 165-167; 1956
        2
1521   Palmer, K.J.; Merrill, R.C.; Ballantyne, M.
        Equilibrium moisture and X-ray diffraction investigations of
        pectinic and pectic acids.
        Journal of the American Chemical Society 70 (2) 570-577; 1948
        2,5
1522   Palmer, K.J.; Shaw, T.M.; Ballantyne, M.
        X-ray and moisture equilibrium investigation of sodium
        pectate.
        Journal of Polymer Science 2 (3) 318-328; 1947
        2,5
1523   Palmia, F.
        Determination of water activity (aw) in cured raw hams as a
        function of water and salt contents.(Orig. ital.)
        Industria Conserve 57 (2) 69-72; 1982
        2,5
1524   Palmia, F.; Pezzani, G.; Raczynski, R.G.
        Determinazione dell'attività dell'aqua (aw) d'insaccatti crudi
        in funzione del contenuto di aqua e di sale.
        Industria Conserve 54 (4) 308-312; 1979
        2,5
1525   Palnitkar, M.P.
        Thermodynamic characteristics of low and intermediate moisture
        foods.
        Ph.D. Thesis, State Unversity, East Lansing, Michigan, USA;
        1971
        2,4

1526 Palnitkar, M.P.; Heldman, D.R.
Equilibrium moisture characteristics of freeze-dried beef components.
Journal of Food Science 36 (7) 1015-1018; 1971
3,5

1527 Panchenkov, G.M.; Tsabek, L.K.; Rozen, I.V.
Dynamics of sorption under equilibrium conditions at a phase boundary characterized by a linear isotherm.(Orig. russ.)
Zurnal Fiziceskoj Chimii 43 (2) 532-536; 1969
2

1528 Pande, A.
Techniques for the measurement of moisture in biological materials.
Laboratory Practice 20 (2) 117-120, 131; 1971
3

1529 Pande, A.
Handbook of moisture determination and control. Principles, techniques, applications. Vol. 3.
Marcel Dekker Inc., New York; 1975
1,3

1530 Pap, L.
Hygroscopicity of wheat.
Cereal Chemistry 8 (-) 200-206; 1931
3,5

1531 Papanicolaou, D.; Rigaud, J.; Sauvageot, F.; Dubois, P.; Simatos, D.
Behaviour of some volatile compounds during storage of orange juice powder with low and intermediate moisture contents.
Journal of Food Technology 13 (-) 511-519; 1978
4

1532 Parasiewicz-Kaczmarska, J.
Preparation of aluminum silicates.
II. Sorption of water vapor, structure, and properties of the samples prepared from silicic acid sol and aluminum chloride.(Orig. pol.)
Zeszyty Naukowe Uniwersytetu Jagiellonskiego Pr. chem.
14 (-) 237-256; 1969
5

1533 Parducci, L.; Duckworth, R.
Differential thermal analysis of frozen food systems.
II. Micro-scale studies on eggwhite, cod and celery.
Journal of Food Technology 7 (4) 423-430; 1972
2

1534 Park, S.W.
A theoretical investigation of sorption kinetic models and its application to the sorption kinetics of water vapor and carbon tetrachloride vapor by cereal grains.
Ph.D. Thesis, Kansas State University, Manhattan, Kansas, USA; 1975
1,2,3,5

1535 Park, S.W.; Chung, D.S.; Watson, C.A.
Adsorption kinetics of water vapour by yellow corn.
I. Analysis of kinetic data for sound corn.
Cereal Chemistry 48 (1) 14-22; 1971
2

1536 Parkhomenko, V.V.; Tretinnik, V.Y.; Kruglitskii, N.N.; Ovcharenko, F.D.; Novotorov, A.S.
Thermodynamic treatment of water vapor sorption by humic acid and humates.(Orig. russ.)
Kolloidnyi Zurnal 31 (2) 269-275; 1969
2,5

1537 Pasch, J.H.; Elbe, J.H. von
Betanine degradation as influenced by water activity.
Journal of Food Science 40 (6) 1145-1146; 1975
4

1538 Passy, N.; Mannheim, C.H.
Flow properties and water sorption of food powders.
II. Egg powders.
Lebensmittel-Wissenschaft und -Technologie 15 (4) 222-225; 1982
2,4,5

1539 Patel, K.; Nickerson, A.
Influence of sucrose, glucose and lactose on loss of water
from solutions.
Journal of Food Science 36 (3) 495-497; 1971
2,4

1540 Pauling, L.
The adsorption of water by proteins.
Journal of the American Chemical Society 67 (4) 555-557; 1945
2

1541 Pavey, R.; Schack, W.
Formulation of intermediate moisture bitesize food cubes.
U.S.A.F. School of Aerospace Medicine, (Technical Report on
Contract F 41609-67-C-0054); 1969
4

1542 Pawsey, R.; Davies, R.
The safety of intermediate moisture foods with respect to
Staphylococcus aureus.
In: Davies, R.; Birch, G.G.; Parker, K.J. (Edts.).
Intermediate Moisture Foods. pp 182-202.
Applied Science Publ., London; 1976
4

1543 Pelaez, J.; Karel, M
Development and stability of intermediate moisture tortillas.
Journal of Food Processing and Preservation 4(1/2) 51-66; 1980
4,5

1544 Peleg, M; Mannheim, C.H.
The mechanism of caking of powdered onion.
Journal of Food Processing and Preservation 1 (-) 3-11; 1977
4,5

1545 Peppas, N.A.; Khanna, R.
Mathematical analysis of transport properties of polymer films
for food packaging.
II. Generalized water vapor models.
Polymer Engineering and Science 20 (17) 1147-1156; 1980
2

1546 Peri, C.; De Cesari, L.
Thermodynamics of water sorption on Sacc. Cerevisiae and cell
viability during spray-drying.
Lebensmittel-Wissenschaft und -Technologie 7 (2) 76-81; 1974
2,5

1547 Pfennig, H.
Dampfdruck einer gesättigten, wäßrigen MgCl2-Lösung in
Abhängigkeit von der Temperatur.
Naturwissenschaften 49 (4) 81; 1962
3

1548 Pfundt, L.A.J.; Spreekens, K.J.A. van
The microbiological quality of bakery products with a high
water activity.(Orig. neth.)
Voedingsmiddelentechnologie 12 (17) 32-34; 1979
4

1549 Pichler, H.J.
      Sorptionsisothermen für Getreide und Raps.
      Landtechnische Forschung 6 (-) 47-52; 1956
      3,5
1550 Pierce, C.
      Effects of interparticle condensation on heats of adsorption
      and isotherms of powder samples.
      Journal of Physical Chemistry 63 (-) 1076-1079; 1959
      2
1551 Pierce, C.
      The Frenkel-Halsey-Hill adsorption isotherm and capillary
      condensation.
      Journal of Physical Chemistry 64 (-) 1184; 1960
      2
1552 Pierce, C.
      Universal nitrogen isotherm.
      Journal of Physical Chemistry 72 (10) 3673-3676; 1968
      5
1553 Pierce, C.
      The Hill equation for adsorption on uniform surfaces.
      Journal of Physical Chemistry 72 (6) 1955-1959; 1968
      2
1554 Pierce, C.; Ewing, B.
      Physical adsorption in the multilayer region on heterogeneous
      and homogeneous surfaces.
      Journal of the American Chemical Society 84 (-) 4070; 1962
      2
1555 Pierce, C.; Smith, R.N.
      Adsorption - desorption hysteresis in relations to capillarity
      of adsorbents.
      Journal of Physical Chemistry 54 (-) 784-794; 1950
      2
1556 Pierce, C.; Smith, R.N.
      Heats of adsorption.
      IV. Entropy changes in adsorption.
      Journal of Physical Chemistry 54 (-) 795; 1950
      2
1557 Pietsch, W.; Rumpf, H.
      Haftkraft, Kapillardruck, Flüssigkeitsvolumen und Grenzwinkel
      einer Flüssigkeitsbrücke zwischen zwei Kugeln.
      Chemie-Ingenieur-Technik 39 (15) 885-893; 1967
      2
1558 Pilosof, A.M.R.; Bartholomai, G.B.; Chirife, J.; Boquet, R.
      Effect of heat treatment on sorption isotherms and solubility
      of flour and protein isolates from bean phaseolus vulgaris.
      Journal of Food Science 47 (4) 1288-1290; 1982
      5
1559 Pinaga, F.; Lafuente, G.
      Horchata en polvo.
      I. Humedades de equilibrio de la horchata liofilizada.
      Revista de Agroquimica y Tecnologia de Alimentos 5 (1) 99;
      1965
      5
1560 Pineri, M.H.; Escourbes, M.; Roche, G.
      Water-collagen interactions: Calorimetric and mechanical
      experiments.
      Biopolymers 17 (-) 2799-2815; 1978
      2

1561 Pintauro, N.
      Intermediate moisture foods.
      In: Pintauro, N.
      Food Additives to extend Shelf Life.
      Park Ridge, Noyes Data Corp.; 1974
      4
1562 Pitt, J.I.
      Xerophilic fungi and the spoilage of foods of plant origin.
      In: Duckworth, R.B. (Ed.).
      Water Relations of Foods. pp 273-307.
      Academic Press, London; 1975
      4
1563 Pixton, S.W.
      The importance of moisture and equilibrium relative humidity
      in stored products.
      Tropical Stored Product Information 43 (1) 16-29; 1982
      5
1564 Pixton, S.W.; Griffiths, H.J.
      Diffusion of moisture through grain.
      Journal of Stored Products Research 7 (3) 133-152; 1971
      2
1565 Pixton, S.W.; Henderson, S.
      The moisture content-equilibrium relative humidity
      relationships of five varieties of Canadian wheat and of Candle
      rapeseed at different temperatures.
      Journal of Stored Products Research 17 (4) 187-190; 1981
      5
1566 Pixton, S.W.; Henderson, S.
      The moisture content-equilibrium relative humidity
      relationship of malt.
      Journal of Stored Products Research 17 (4) 191-195; 1981
      5
1567 Pixton, S.W.; Warburton, S.
      The time required for conditioning grain to equilibrium with
      specific relative humidities.
      Journal of Stored Products Research 4 (-) 261-265; 1968
      2,5
1568 Pixton, S.W.; Warburton, S.
      Moisture content / relative humidity equilibrium of some
      cereal grains at different temperature.
      Journal of Stored Products Research 6 (-) 283-293; 1971
      5
1569 Pixton, S.W.; Warburton, S.J.
      Moisture content / relative humidity equilibrium at different
      temperatures of some oil seeds of economic importance.
      Journal of Stored Products Research 7 (-) 261-269; 1971
      5
1570 Pixton, S.W.; Warburton, S.
      Determination of moisture content and equilibrium relative
      humidity of dried fruit-sultanas.
      Journal of Stored Products Research 8 (-) 263-270; 1973
      5
1571 Pixton, S.W.; Warburton, S.
      The influence of the method used for moisture adjustment on
      the equilibrium relative humidity of stored products.
      Journal of Stored Products Research 9 (-) 189-197; 1973
      3,5
1572 Pixton, S.W.; Warburton, S.
      The moisture content/equilibirum relative humidity
      relationship of macaroni.
      Journal of Stored Products Research 9 (-) 247-251; 1973
      5

-144-

1573 Pixton, S.W.; Warburton, S.
      The moisture content / equilibrium relative humidity
      relationship of soya meal.
      Journal of Stored Products Research 11 (-) 249-251; 1975
      5
1574 Pixton, S.W.; Warburton, S.
      The moisture content / equilibrium relative humidity
      relationship of dried yeast product.
      Journal of Stored Products Research 13 (1) 35-37; 1977
      5
1575 Pixton, S.W.; Warburton, S.
      The moisture content / equilibrium relative humidity
      relationship and oil composition of rapeseed.
      Journal of Stored Products Research 13 (-) 77-81; 1977
      5

      Pixton, S.W. see also 2197-2199

1576 Plett, E.A.
      Wasseraktivität und Aromaerhaltung.
      In: Loncin, M. (Ed.).
      Hochschulkurs Ausgewählte Themen der
      Lebensmittelverfahrenstechnik "Wasseraktivität".
      Institut für Lebensmittelverfahrenstechnik Universität
      Karlsruhe; 1980
      4
1577 Plitman, M.; Park, Y.; Gomez, R.; Sinskey, A.J.
      Viability of Staphylococcus aureus in intermediate moisture
      meats.
      Journal of Food Science 38 (6) 1004-1008; 1973
      4,5
1578 Poersch, W.
      Sorptionsisothermen - Ihre Ermittlung und Auswertung.
      Die Stärke 15 (11) 405-412; 1963
      1,3,5
1579 Poisson, J.; Guilbot, A.
      Sorptionsisotherme von Weizen.(Orig. fr.)
      La Meunerie Francaise 163 (-) 11; 1963
      5
1580 Pomeranz, Y.
      Biochemical, functional and nutritive changes during storage.
      In: Christensen, C.M (Ed.).
      Storage of cereal grains and their products. pp 56-114.
      American Association of Cereal Chemists, Incorporated,
      St. Paul, Minnesota; 1974
      4,5
1581 Potter, N.N.
      Intermediate moisture foods: principles and technology.
      Food Product Development 4 (7) 38-48; 1970
      4
1582 Potthast, K.
      Einfluß der Wasseraktivität auf enzymatische Vorgänge in
      gefriergetrocknetem Muskelfleisch.
      Dissertation, Universität Münster; 1972
      4
1583 Potthast, K.
      Haltbarkeit, Herstellung und Verarbeitung von
      gefriergetrocknetem Fleisch.
      Die Fleischwirtschaft 57 (9) 1618-1624; 1977
      4,5

1584 Potthast, K.
Haltbarkeit und Verarbeitung von gefriergetrocknetem Fleisch.
Gordian 77 (6) 158-161; 1977
1,4,5

1585 Potthast, K.
Wasseraktivität, Gefriertrocknung und Rohwurstherstellung.
Die Fleischwirtschaft 58 (11) 1792-1796; 1978
1,4

1586 Potthast, K.; Acker, L.; Hamm, R.
Einfluß der Wasseraktivität auf enzymatische Veränderungen in
gefriergetrocknetem Muskelfleisch.
Zeitschrift für Lebensmittel-Untersuchung und -Forschung
165 (1) 15-17; 1977
4

1587 Potthast, K.; Acker, L.; Hamm, R.
Einfluß der Wasseraktivität auf enzymatische Veränderungen in
gefriergetrocknetem Muskelfleisch.
IV. Änderung der Aktivität glykolytischer Enzyme während der
Lagerung.
Zeitschrift für Lebensmittel-Untersuchung und Forschung
165 (1) 18-20; 1977
4

1588 Potthast, K.; Hamm, R.; Acker, L.
Enzymic reactions in low moisture foods.
In: Duckworth, R.B. (Ed.).
Water Relations of Foods. pp 365-377.
Academic Press, London; 1975
4

1589 Potthast, K.; Hamm, R.; Acker, L.
Einfluß der Wasseraktivität auf enzymatische Veränderungen in
gefriergetrocknetem Muskelfleisch.
II. Reaktionen der Kohlenhydrate.
Zeitschrift für Lebensmittel-Untersuchung und -Forschung
162 (2) 139-148; 1976
4

1590 Poulsen, K.P.; Lindelov, F.
Acceleration of chemical reactions due to freezing.
In: Rockland, L.B.; Stewart, G.F. (Edts.).
Water Activity: Influences on Foodquality. pp 651-678.
Academic Press, New York; 1981
4

1591 Poulter, R.G.; Doe, P.E.; Olley, J.
Isohalic sorption isotherms.
II. Use in the prediction of storage life of dried salted
fish.
Journal of Food Technology 17 (2) 201-210; 1982
2,4

1592 Pouncy, A.E.; Summers, B.C.L.
The micro-measurement of relative humidity for the control of
osmophilic yeasts in confectionery products.
Journal of the Society of Chemical Industry (London)
58 (-) 162-165; 1939
3,4

1593 Powers, H.E.C.
Refined sugar moisture.
International Sugar Journal 56 (-) 314-315; 1954
5

1594 Powers, M.J.; Jackson, R.; Kilpatrick, P.W.; Brekke, J.E.;
Sanshuck, D.W.
Low-moisture, red sour pitted cherries.
Food Technology 12 (10) 610-613; 1958
5

1595 Prabhakar, J.V.
Influence of water activity on secondary reactions of
autoxidation of lipids.
In: Proc. First Indian Convention of Food Scientists
and Technologists (see FSTA 1979 11 12A871).
No. 1.10, pp 9-10; 1979
4

1596 Prabhakar, J.V.; Amla, B.L.
Influence of water activity on the formation of monocarbonyl
compounds in oxidizing walnut oils.
Journal of Food Science 43 (-) 1839-1843; 1978
4

1597 Prado-Filho, L.G.
Die thermische Abtötung von Sporen von Bacillus subtilis var.
niger ATCC 9372 in Gasphasen mit einer Wasseraktivität unter
1.0.
Lebensmittel-Wissenschaft und -Technologie 8 (1) 29-33; 1975
4

1598 Prado Filho, L.G.; Kiefer, P.
Versuchsanlage zur thermischen Behandlung von Bakteriensporen
in Gasen mit einer Wasseraktivität <1.0.
Lebensmittel-Wissenschaft und -Technologie 7(4) 245-246; 1974
4

1599 Pratap, V.; Singh, B.P.N.; Narain, M.
Equilibrium moisture content of some flours.
Journal of Food Science and Technology, India 19 (4) 153-158;
1982
2,5

1600 Price, C.A.
Note on the surface contamination of glass specimens held at
constant relative humidity over saturated salt solutions.
Glass Technology 21 (6) 306; 1980
3

1601 Prior, B.A.
Measurement of the water activity in food: A review.
Journal of Food Protection 42 (8) 668-674; 1979
1,3

1602 Prior, B.A.; Casaleggio, C.; Vuuren, J.J. van
Psychrometric determination of water activity in the high aw
range.
Journal of Food Protection 40 (8) 537-539; 1977
3

1603 Prokhorova, A.; Kretovich, V.
Enzymic processes in stored dry vegetable materials.(Orig.russ.
Biokhimiya Zerna, Sbornik 4 (-) 132-137; 1958
4

1604 Pruthi, J.S.; Singh, L.J.; Girdhavi, L.
The equilibrium relative humidity of garlic powder.
Journal of the Science of Food and Agriculture 10 (-) 359-361;
1959
5

1605 Punnett, P.W.
Effects of relative humidity on coffee.
Tea and Coffee Trade Journal 117 (6) 24-28; 1959
5

1606 Puri, B.R.; Toteja, K.K.; Malik, R.C.
Physicochemical properties of caseins.
II. Sorption of water vapor by caseins.
Journal of the Indian Chemical Society 46 (6) 554-56 ; 1969
2,5

1607 Purr, A.
Zum Ablauf chemischer Veränderungen in wasserarmen
Lebensmitteln.
I. Der enzymatische Abbau der Fette bei niedrigen
Wasserdampfpartialdrücken.
Fette, Seifen, Anstrichmittel 68 (2) 145-154; 1966
4,5
1608 Purwadaria, H.K.; Heldman, D.R.; Kirk, J.R.
Computer simulation of vitamin degradation in a dry model food
system during storage.
Journal of Food Process Engineering 3 (1) 7-28; 1980
4
1609 Puski, G.
Modification of functional properties of soy proteins by
proteolytic enzyme treatment.
Cereal Chemistry 52 (-) 655-664; 1975
4
1610 Putranon, R.; Bowrey, R.G.; Eccleston, J.
Sorption isotherms for two cultivars of paddy rice grown in
Australia.
Food Technology in Australia 31 (12) 510-515; 1979
3,5
1611 Putranon, R.; Bowrey, R.G.; Fowler, R.T.;
The effect of moisture content on the heat of sorption and
specific heat of Australian paddy rice.
Food Technology in Australia 32 (2) 56-59; 1980
2,5
1612 Quast, D.G.; Karel, M.; Rand, W.M.
Development of a mathematical model for oxidation of potato
chips as a function of oxygen pressure, extent of oxidation
and equilibrium relative humidity.
Journal of Food Science 37 (5) 673-678; 1972
4
1613 Quast, D.G.; Teixeira Neto, R.O.
Atividade de agua emalguns alimentos de teor intermediario de
umidade.
Coletanea do Instituto de Tecnologia de Alimentos
6 (1) 203-232; 1975
4
1614 Quast, D.G.; Teixeira Neto, R.O.
Moisture problems of foods in tropical climates.
Food Technology 30 (5) 98, 102, 104-105; 1976
4,5
1615 Quinn, F.C.
Total moisture content approach oversimplifies processing
complexity.
Candy Industry and Confectioners' Journal 134 (12) 16, 18;
1970
1
1616 Quinn, M.R.; Beuchat, L.R.
Functional property changes resulting from fungal fermentation
of peanut flour.
Journal of Food Science 40 (-) 475-478; 1975
5
1617 Radtke, R.; Heiss, R.
Über das Lagerungsverhalten von Haselnüssen türkischer
Provenienz.
Süßwaren 15 (3/4) 103-106, 137-142; 1971
4,5

1618 Rahman, A.R.; Schafer, G.; Prell, P.; Westcott, D.E.
Non-reversible compression of intermediate moisture fruit
bars.
United States Army, Natick Laboratories, Natick, Massachusetts
01760; Technical Report TR 71-60-FL
4

1619 Rakowski, A.
Zur Kenntnis der Adsorption.
I. Chemische Hysteresis der Stärke.
Zeitschrift für Chemie und Industrie der Kolloide
(Kolloid-Zeitschrift) 9 (2) 225-230; 1911
2,5

1620 Rakowski, A.
Zur Kenntnis der Adsorption.
V. Chemische Hysteresis der Stärke.
Kolloidnyi Zhurnal 11 (-) 269-272; 1912
2

1621 Ramakrishnan, T.V.; Francis, F.J.
Stability of carotenoids in model aqueous systems.
Journal of Foodquality 2 (-) 177-189; 1979
4,5

1622 Ramanuja, M.N.; Jayaraman, K.S.
Proc. First Indian Convention of Food Scientists
and Technologists.
Summaries of papers and discussions, June 1978, p. 86,
No. 8.4; Mysore, India; 1979
4

1623 Ramanuja, M.N.; Jayaraman, K.S.
Studies on the preparation and storage stability of
intermediate moisture banana.
Journal of Food Science and Technology 17 (4) 183-186; 1980
4

1624 Ramirez-Martinez, J.R.; Levi, A.; Padua, H.; Bakal, A.
Astringency in an intermediate moisture banana product.
Journal of Food Science 42 (5) 1201-1203, 1217; 1977
4

1625 Ramstad, P.E.; Geddes, W.F.
The respiration and storage behaviour of soybeans.
Agr. Exp. Sta. Tech. Bul. 156 University of Minnesota; 1942
5

1626 Ranganna, S.; Srinivasan, B.
Ascorbic acid-amino acid browning in relation to water
activity.
In: Proc. First Indian Convention of Food Scientists
and Technologists. No. 1.1, p. 2; 1979
4

1627 Rangaswamy, J.R.
Observations on the sorption of water vapour by rice and
sorghum.
Journal of Food Science and Technology, India 10 (2) 59-61;
1973
5

1628 Rao, G.N.S.; Rao, K.S.; Rao, B.S.
Hysteresis in sorption.
XV. Hysteresis in the sorption of water on casein, egg albumin
and gelatin.
Proc. of the Indian Academy of Sciences
A 25 (-) 221-228; 1947
2,5

1629 Rao, K.S.
Elasticity of organo-gels in relation to hysteresis in sorption.
Current Science 9 (1) 19-21; 1940
2,4,5

1630 Rao, K.S.
Hysteresis in sorption. I-VI.
Journal of Physical Chemistry 45 (-) 500-539; 1941
2,3,5

1631 Rao, K.S.; Das, B.
Sorption-desorption hysteresis of water on gelatin.
Current Science 36 (24) 657-658; 1967
5

1632 Rao, K.S.; Das, B.
Varietal differences in gelatin, egg-albumin and casein in relation to sorption-desorption hysteresis with water.
Journal of Physical Chemistry 72 (4) 1223-1230; 1968
2,5

1633 Rao, K.S.; Das, B.
Sorption-desorption hysteresis in glassy silica gel and fibrous silica gel (Santocel C) with water, carbon tetrachloride, normal aliphatic alcohols.
Journal of Colloid and Interface Science 32 (1) 24-32; 1970
2,5

1634 Rasekh, J.G.; Hale, M.B.; Goldmintz, D.; Sidwell, V.D.
Using fish in intermediate-moisture low-bacteria pet foods.
Food Product Development (10) 66-69; 1976
1,4

1635 Rasekh, J.G.; Stillings, B.R.; Dubrow, D.L.
Moisture adsorption of fish protein concentrate at various relative humidities and temperatures.
Journal of Food Science 36 (4) 705-707; 1971
5

1636 Razga, Z.
Der Zusammenhang zwischen Wassergehalt und "Hydratur" des Hopfens.
Brauwissenschaft 16 (1) 12-16; 1963
3,5

1637 Reade, M.G.
Hygrometry of cocoa and chocolate products.
Revue Internationale de la Chocolaterie 24 (2) 34-45; 1969
3,4,5

1638 Reade, M.G.
Hygrometry of confectionary products.
In: Relative humidity in the food industry.
B.F.M.I.R.A.-Symposium Proc. No. 4, pp 37-43, Leatherhead, London; 1969
3,4,5

1639 Rees, W.H.
The heat of absorption of water by cellulose.
Journal of Textile Institute 39 (-) 1351-1367; 1948
2

1640 Rehacek, J.; Sozzi, T.; Studer, P.
Effect of water activity on the development of lactic acid bacteria and yeast utilized in the food industry.
Milchwissenschaft 3A (3) 151-154; 1982
4

1641 Reichert, E.; Stiebing, A.
Herstellung von längerfristig haltbaren Leberwurstkonserven durch Pasteurisieren infolge aw-Wertsenkung.
Die Fleischwirtschaft 57 (5) 910-916, 921; 1977
1,4

1642 Reid, D.S.
      Water activity concepts in intermediate moisture foods.
      In: Davies, R.; Birch, G.G.; Parker, K.J. (Edts.).
      Intermediate Moisture Foods. pp 54-65.
      Applied Science Publ., London; 1976
      4,5

1643 Reiss, J.
      Mycotoxins in foodstuffs.
      XII. The influence of the water activity (aw) of cakes on the
      growth of moulds and the formation of mycotoxins.
      Zeitschrift für Lebensmittel-Untersuchung und -Forschung
      167 (6) 419-422; 1978
      2

1644 Resnik, S.; Chirife, J.
      Effect of moisture content and temperature on some aspects of
      non-enzymatic browning in dehydrated apple.
      Journal of Food Science 44 (2) 601-605; 1979
      4,5

1645 Resnik, S.L.; Plett, E.A.; Loncin, M.
      Untersuchungen zur nicht-enzymatischen Bräunungsreaktion in
      flüssigen Modellsystemen.
      Zeitschrift für Lebensmittel-Technologie und
      -Verfahrenstechnik 32 (5) 213-216; 1981
      4

1646 Rey, D.K.; Labuza, T.P.
      Characterization of the effect of solutes on the water-binding
      and gel strength properties of carrageenan.
      Journal of Food Science 46 (3) 786-789, 793; 1981
      2

1647 Reyerson, L.H.; Hnojewyj, W.S.
      The sorption of H2O and D2O vapors by lyophilized
      beta-lactoglobulin and the deuterium-exchange effect.
      Journal of Physical Chemistry 64 (-) 811-815; 1960
      2,5

1648 Rheinbothe, H.
      Ein Beitrag zur Bestimmung der Luftfeuchtigkeit bei höheren
      Temperaturen durch psychrometrische Messungen.
      Dissertation, Thechnische Hochschule Karl-Marx-Stadt; 1963
      3

1649 Rhodes, E.R.
      Phosphate sorption isotherms for some Sierra Leone soils.
      Journal of the Science of Food and Agriculture 26 (-) 895-902;
      1975
      2

1650 Richards, E.
      Non-enzymic browning: the reaction between D-glucose and
      glycine in the dry state.
      Biochemical Journal 64 (4) 639-644; 1956
      4

1651 Richardson, G.M.; Malthus, R.S.
      Salts for static control of humidity at relatively low levels.
      Journal of Applied Chemistry 5 (10) 557-567; 1955
      3

1652 Riedel, L.
      Zum Problem des gebundenen Wassers in Fleisch.
      Kältetechnik 13 (3) 122-128; 1961
      2,5

1653 Riedel, L.
      Wasser.
      In: Schormüller, J. (Ed.).
      Handbuch der Lebensmittelchemie Bd. 1, pp 100-122. Springer
      Verlag, Berlin-Heidelberg; 1965
      1,2,5

1654 Riedel, L.
Kalorimetrische Bestimmung der Hydratationswärme von
Lebensmitteln.
Chemie Mikrobiologie Technologie der Lebensmittel
5 (4) 97-101; 1976
2

1655 Riemann, H.
Effect of water activity on the heat resistance of salmonella
in "dry" materials.
Applied Microbiology 16 (10) 1621; 1968
4

1656 Riemer, J.; Karel, M.
Shelf life studies of vitamin C during food storage:
prediction of L-ascorbic acid retention in dehydrated tomato
juice.
Journal of Food Processing and Preservation 1 (4) 293-312;
1977
4

1657 Riet, W.B. van der
Water sorption isotherms of beef biltong and their use in
predicting critical moisture contents for biltong storage.
South African Food Review 3 (6) 93-94, 96; 1976
4,5

1658 Rigg, W.J.
Measurement of the permeability of chilled meat packaging film
under conditions of high humidity.
Journal of Food Technology 14 (2) 149-155; 1979
4

1659 Rigler, F.
Einfluß der relativen Luftfeuchtigkeit auf die Haltbarkeit von
Rohwürsten.
Archiv für Lebensmittelhygiene 14 (8) 185-186; 1963
4

1660 Ripperger, S.; Germerdonk, R.
Das Adsorptionsgleichgewicht von unterschiedlich mit Wasser
mischbaren organischen Luftschadstoffen und von Wasser an
Aktivkohle.
Chemie-Ingenieur-Technik 54 (12) 1204-1205; 1982
5

1661 Robens, E.
Gravimetric sorption measurements.
Laboratory Practice 18 (3) 292-299, 314; 1969
3,5

1662 Robens, E.; Sandstede, G.
Apparatur zur automatischen Wägung adsorbierter Gase bei
0.01 - 1000 Torr.
Chemie-Ingenieur-Technik 40 (19) 957-960; 1968
3

1663 Robens, E.; Walter, G.
Oberflächen- und Porenbestimmung durch Sorptionsmessung.
CZ-Chemie-Technik 1 (3) 127-129; 1972
3

1664 Roberts, B.F.
A procedure for estimating pore volume and area distributions
from sorption isotherms.
Journal of Colloid and Interface Science 23 (-) 266-273; 1967
3

1665 Roberts, T.A.; Smart, J.L.
Control of clostridia by water activity and related factors.
In: Davies, R.; Birch, G.G.; Parker, K.J. (Edts.).
Intermediate Moisture Foods. pp 203-214.
Applied Science Publ., London; 1976
4

1666   Robertson, D.W.; Lute, A.M.; Gardner, R.
        Effect of relative humidity on viability, moisture content and
        respiration of wheat, oats and barley seeds in storage.
        Journal of Agricultural Research 59 (-) 281-291; 1939
        4,5

1667   Robinson, R.A.
        Standard solutions for humidity control at 25 °C.
        Transactions of the Faraday Society 41 (-) 756-758; 1945
        3

1668   Robinson, R. A.
        Activity coefficients of sodium chloride and potassium
        chloride in mixed aqueous solutions at 25 °C.
        Journal of Physical Chemistry 65 (-) 662-667; 1961
        3

1669   Robinson, R.A.; Bower, V.E.
        An additivity rule for the vapor pressure lowering of aqueous
        solutions.
        Journal of Research of the National Bureau of Standards
        69 A (4) 365-367; 1965
        3

1670   Robinson, R.A.; Sinclair, D.A.
        The activity of the alkali chlorides and of lithium iodide in
        aqueous solutions from vapor pressure measurements.
        Journal of the American Chemical Society 56 (-) 1830-1835; 1934
        3

1671   Robinson, R.A.; Stokes, R.H.
        Activity coefficients in aqueous solutions of sucrose,
        mannitol, and their mixtures at 25 °C.
        Journal of Physical Chemistry 65 (-) 1954-1958; 1961
        3

1672   Robinson, R.A.; Stokes, R.H.
        Electrolyte solutions.
        Butterworth, London; 1965
        2,3

1673   Robson, J.N.
        Some introductory thoughts on intermediate moisture foods.
        In: Davies, R.; Birch, G.G.; Parker, K.J. (Edts.).
        Intermediate Moisture Foods. pp 32-42.
        Applied Science Publ., London; 1976
        4

1674   Rochester, C.H.; Westerman, A.V.
        Gravimetric study of the sorption of water vapour by bovine
        serum albumin.
        Journal of the Chemical Society, Faraday Transactions I,
        72 (-) 2498-2504; 1976
        5

1675   Rochester, C.H.; Westerman, A.V.
        Sorption of water vapour by poly-L-glutamic acid,
        poly-L-lysine and their salts and some chemically modified
        derivatives.
        Journal of the Chemical Society, Faraday Transactions I,
        72 (-) 2753-2768; 1976
        5

1676   Rockland, L.B.
        A new treatment of hygroscopic equilibria application to
        walnuts (Juglans regia) and other foods.
        Food Research 22 (6) 604-628; 1957
        1,2,3,5

1677   Rockland, L.B.
        Saturated salt solutions for static control of relative
        humidity between 5 and 40 °C.
        Analytical Chemistry 32 (10) 1375-1376; 1960
        3

1678 Rockland, L.B.
      Water activity and storage stability.
      Food Technology 23 (10) 1241-1248; 1969
      1,2,5
1679 Rockland, L.B.; Nishi, S.K.
      Influence of water activity on food product quality and
      stability.
      Food Technology 34 (4) 42-51, 59; 1980
      1,2,3
1680 Rockland, L.B.; Stewart, G.F. (Edts.).
      Water activity. Influences on Foodquality. A treatise on the
      inluence of bound and free water on the quality and stability
      of foods and other natural products.
      Academic Press, New York; 1981
      1,2,3,4,5
1681 Rockland, L.B.; Swarthout, D.M.; Johnson, R.A.
      Studies on english (Persian) walnuts, Juglans regia.
      III. Stabilization of kernels.
      Food Technology 15 (3) 112-116; 1961
      4
1682 Rodriguez-Arias, J.H.
      Desorption isotherms and drying rates of shelled corn in the
      temperature range of 40 to 140 °F.
      Ph.D. Thesis, Department of Agricultural Engineering,
      Michigan State University, East Lansing; 1956
      2,5
1683 Rodriguez-Arias, J.H.; Hall, C.W.; Bakker-Arkema, F.
      Heat of vaporization for shelled corn.
      Cereal Chemistry 40 (6) 676-683; 1963
      2,5
1684 Rödel, W.
      Die Wasseraktivität, ein wichtiger Faktor bei der Herstellung
      von Fleischwaren.
      Die Fleischwirtschaft 52 (12) 1550-1552; 1972
      1,4
1685 Rödel, W.
      Messung der Wasseraktivität unter Praxisbedingungen.
      Die Fleischwirtschaft 53 (1) 27-31; 1973
      1,3
1686 Rödel, W.
      Der aw-Wert - ein wichtiger Faktor für die Praxis der Fleisch-
      waren-Herstellung.
      Fleischerei 24 (6) 44-46; 1973
      1,4
1687 Rödel, W.
      Einstufung von Fleischerzeugnissen in leicht verderbliche,
      verderbliche und lagerfähige Produkte aufgrund des pH-Wertes
      und aw-Wertes.
      Dissertation, Freie Universität Berlin; 1975
      1,4
1688 Rödel, W.
      Feststellung und Bedeutung des aw-Wertes von Rohwurst für die
      Lebensmittelüberwachung.
      Die Fleischwirtschaft 55 (4) 498-499; 1975
      1,3,4
1689 Rödel, W.; Herzog, H.; Leistner, L.
      Wasseraktivitäts-Toleranz von lebensmittelhygienisch wichtigen
      Keimarten der Gattung Vibrio.
      Die Fleischwirtschaft 53 (9) 1301-1303; 1973
      4

1690 Rödel, W.; Krispien, K.
Der Einfluß von Kühl- und Gefriertemperaturen auf die
Wasseraktivität (aw-Wert) von Fleisch und Fleischerzeugnissen.
Die Fleischwirtschaft 57 (10) 1863-1867; 1977
1,4,5
1691 Rödel, W.; Krispien, K.; Hofmann, G.
Bedeutung und Messung der Oberflächenwasseraktivität von
Fleisch und Fleischerzeugnissen.
Die Fleischwirtschaft 60 (10) 1840-1844; 1980
1,3
1692 Rödel, W.; Krispien, K.; Leistner, L.
Messung der Wasseraktivität (aw-Wert) von Fleisch und
Fleischerzeugnissen.
Die Fleischwirtschaft 59 (6) 831-836; 1979
1,3
1693 Rödel, W.; Krispien, K.; Leistner, L.
Die Wasseraktivität von Fetten tierischer Herkunft.
Die Fleischwirtschaft 60 (4) 642-650; 1980
4,5
1694 Rödel, W.; Leistner, L.
Ein einfacher aw-Wert-Messer für die Praxis.
Die Fleischwirtschaft 51 (12) 1800-1802; 1971
1,3
1695 Rödel, W.; Leistner, L.
Messung der Wasseraktivität (aw-Wert) von Fleisch und
Fleischwaren mit einem Taupunkt Hygrometer.
Die Fleischwirtschaft 52 (11) 1461-1462; 1972
3
1696 Rödel, W.; Leistner, L.
Möglichkeiten und Grenzen der Fleischwarenherstellung als
"Intermediate Moisture Foods".
Mitteilungsblatt der Bundesanstalt für Fleischforschung,
Kulmbach, p. 2613; 1975
4
1697 Rödel, W.; Leistner, L.
Kritische Würdigung der Einstufung von Fleischerzeugnissen
nach dem aw-Wert und pH-Wert.
Die Fleischwirtschaft 62 (3) 288-291; 1982
4,5
1698 Rödel, W.; Ponert, H.; Leistner, L.
Verbesserter aw-Wert-Messer zur Bestimmung der Wasseraktivität
(aw-Wert) von Fleisch und Fleischwaren.
Die Fleischwirtschaft 55 (4) 557-558; 1975
3
1699 Rödel, W.; Ponert, H.; Leistner, L.
Einstufung von Fleischerzeugnissen in leicht verderbliche,
verderbliche und lagerfähige Produkte.
Die Fleischwirtschaft 56 (3) 417-418; 1976
4,5
1700 Rösinger, H.
Ein Gerätesystem zur industriellen Luft- und
Gasfeuchtemessung.
Chemie-Technik 10 (9) 867-871; 1981
3
1701 Rogachev, V.I.; Kislenko, I.I.
Combined methods of decreasing water activity in fruits under
preservation.
Acta Alimentaria Academiae Scientarum Hungaricae
2 (3) 245-250; 1973
3,4

1702 Rogers, M.N.
A small electrical hygrometer for microclimate measurements.
In: Wexler, A.; Ruskin, R.E. (Edts.).
Humidity and Moisture. Measurement and Control in Science and
Industry. Vol. 1, pp 302-309. Reinhold Publ. Corporation,
New York; 1965
3

1703 Rolfe, E.J.
A place for intermediate moisture foods.
In: Davies, R.; Birch, G.G.; Parker, K.J. (Edts.).
Intermediate Moisture Foods. pp 1-3.
Applied Science Publ., London; 1976
4

1704 Roller, S.D.; Anagnastopoulos, G.D.
Accumulation of carbohydrate by Escherichia coli B/r/1 during
growth at low water activity.
Journal of Applied Bacteriology 52 (6) 425-434; 1982
4

1705 Roman, G.N.; Urbicain, M.J.; Rotstein, E.
Moisture equilibrium in apples at several temperatures:
Experimental data and theoretical considerations.
Journal of Food Science 47 (5) 1484-1488, 1507; 1982
2,5

1706 Romo, C.R.; Bartholomai, G.B.
Functional properties of protein isolate from the bean
Phaseolus vulgaris.
Lebensmittel-Wissenschaft und -Technologie 11 (1) 35-37; 1978
2,5

1707 Romney, D.H.
Moisture studies. 2nd Report.
Jamaica Coconut Industry Board Research Department 18-24; 1962
In: Gough, M.C.; Lipiatt, G.A. (Edts.).
Moisture Humidity Equilibria of Tropical Stored Produce
Part II - Oilseeds.
Tropical Stored Products Information 34 (-) 49-61; 1977
5

1708 Roozen, J.P.; Pilnik, W.
Über die Stabilität adsorbierter Enzyme in wasserarmen
Systemen.
I. Die Stabilität von Peroxidase bei 25 °C.
Lebensmittel-Wissenschaft und -Technologie 3 (2) 37-40; 1970
4

1709 Roozen, J.P.; Pilnik, W.
Über die Stabilität adsorbierter Enzyme in wasserarmen
Systemen.
II. Der Einfluß von pH-Verschiebungen auf die Stabilität von
Peroxidase bei 2 °C.
Lebensmittel-Wissenschaft und -Technologie 4 (1) 24-27; 1971
4

1710 Roozen, J.P.; Pilnik, W.
Über die Stabilität adsorbierter Enzyme in wasserarmen
Systemen.
VI. Möglichkeiten zur Inaktivierung von Enzymen in
getrockneten Lebensmitteln durch Bestrahlung.
Besprechung der Literatur und eigene Untersuchungen.
Lebensmittel-Wissenschaft und -Technologie 5 (4) 128-131; 1972
4

1711 Ross, K.D.
Estimation of water activity in intermediate moisture foods.
Food Technology 29 (3) 26-34; 1975
2

1712 Ross, K.D.
Differential scanning calorimetry of nonfreezable water in
solute-macromolecule-water systems.
Journal of Food Science 43 (6) 1812-1815; 1978
2

1713 Ross, K.D.
Definition of bound water by water activity depression.
Lebensmittel-Wissenschaft und -Technologie 12 (3) 172-176;
1979
2

1714 Roth, D.
Amorphisierung bei der Zerkleinerung und Rekristallisation als
Ursachen der Agglomeration von Puderzucker und Verfahren zu
deren Vermeidung.
Dissertation, Technische Hochschule Karlsruhe; 1976
2,3,5

1715 Roth, D.
Das Wasserdampfsorptionsverhalten von Puderzucker.
Zucker 30 (6) 274-284; 1977
3,5

1716 Roth, T.
Verminderung der Trocknungsgeschwindigkeit von Lebensmitteln
durch grenzflächenaktive Stoffe.
Zeitschrift für Lebensmitteltechnologie und Verfahrenstechnik
33 (7) 497-508; 1982
2

1717 Rounsley, R.R.
Multimolecular adsorption equation.
American Institute of Chemical Engineers, Journal
7 (2) 308-311; 1961
2,5

1718 Rouse, P.E. jr.
Diffusion of vapors in films.
Journal of the American Chemical Society 69 (-) 1068-1073;
1947
2,5

1719 Rowen, J.W.; Blaine, R.L.
Sorption of nitrogen and water vapor on textile fibres.
Industrial and Engineering Chemistry 39 (-) 1659-1663; 1947
2,5

1720 Rüb, F.
Feuchtigkeitsmessung - Verfahren und Geräte.
Ernährungswirtschaft 10 (-) 625-630; 1976
3

1721 Rüegg, M.
Berechnung der Wasseraktivität von Schwefelsäurelösungen bei
verschiedenen Temperaturen.
Lebensmittel-Wissenschaft und -Technologie 13 (1) 22-24; 1980
3

1722 Rüegg, M.; Blanc, B.
Thermoanalytische Untersuchungen an Käse.
Schweizerische Milchwirtschaftliche Forschung 1 (-) 9-15; 1972
2

1723 Rüegg, M.; Blanc, B.
Effect of pH on water vapor sorption by caseins.
Journal of Dairy Science 59 (6) 1019-1024; 1976
4,5

1724 Rüegg, M.; Blanc, B.
Beziehungen zwischen Wasseraktivität, Wasser-Sorptionsvermögen
und Zusammensetzung von Käse.
Milchwissenschaft 32 (4) 193-201; 1977
4,5

1725 Rüegg, M.; Blanc, B.
The water activity of honey and related sugar solutions.
Lebensmittel-Wissenschaft und -Technologie 14 (1) 1-6; 1981
5

1726 Rüegg, M.; Blanc, B.
Influence of water activity on the manufacture and aging of
cheese.
In: Rockland, L.B.; Stewart, G.F. (Edts.).
Water Activity: Influences on Foodquality. pp 791-811.
Academic Press, New York; 1981
4

1727 Rüegg, M.; Blanc, B.; Lüscher, M.
Hydration of casein micelles and isotherms of water sorption
of micellar casein isolated from fresh and heat-treated milk.
Journal of Dairy Research 46 (-) 325-328; 1979
2,5

1728 Rüegg, M.; Eberhard, P.; Moor, U.; Flückiger, E.; Blanc, B.
Beziehungen zwischen Teigbeschaffenheit und Zusammensetzung
von Käse.
Schweizerische Milchwirtschaftliche Forschung 9 (-) 3-8; 1980
4

1729 Rüegg, M.; Glättli, H.; Blanc, B.
Einfluß der Wasseraktivität auf Vermehrung und Stoffwechsel
von Propionsäurebakterien.
Schweizerische Milchwirtschaftliche Forschung 5 (-) 119-122;
1976
4

1730 Rüegg, M.; Häni, H.
Infrared spectroscopy of the water vapor sorption process of
caseins.
Biochimica et Biophysica Acta 400 (-) 17-23; 1975
2

1731 Rüegg, M.; Lüscher, M.; Blanc, B.
Hydration of native and rennin coagulated caseins as
determined by differential scanning calorimetry and
gravimetric sorption measurements.
Journal of Dairy Science 57 (4) 387-393; 1974
2,5

1732 Rüegg, M.; Moor, U.; Blanc, B.
Hydration and thermal denaturation of ß-lactoglobulin.
A calorimetric study.
Biochimica et Biophysica Acta 400 (-) 334-342; 1975
2,5

1733 Rüegg, M.; Moor, U.; Lukesch, A.; Blanc, B.
Hydration and thermal stability of alpha-lactalbumin.
A calorimetric study.
In: Applications of Calorimetry in Life Sciences. pp 59-73.
Walter de Gruyter, Berlin, New York; 1977
2,5

1734 Rutherford, A.
Interpretation of sorption and diffusion data in porous
solids.
Industrial and Engineering Chemistry Fundamentals
22 (1) 150-151; 1983
2

1735 Ruthven, D.M.; Derrah, R.I.
Sorption in Davison 5 A Molecular sieves.
Canadian Journal of Chemical Engineering 50 (-) 743-747; 1972
2,5

1736 Ruthven, D.M.; Kumar, R.
      An experimental study of single-component and binary
      adsorption equilibria by a chromatographic method.
      Industrial and Engineering Chemistry Fundamentals
      19 (-) 27-32; 1980
      2
1737 Sabirov, S.M.
      Adsorption equilibriums.(Orig. russ.)
      Doklady Akademie Nauk Uzbekskoj SSR. Taskent 24 (11) 29-31;
      1967
      2
1738 Sahoo, B.
      Study of intermediate moisture pork and its storage stability.
      Ph.D. Thesis, University of Missouri Miss. USA; 1971
      4
1739 Sair, L.; Fetzer, W.R.
      Water sorption by starches.
      Industrial and Engineering Chemistry 36 (3) 205-208; 1944
      2,4,5
1740 Sair, L.; Fetzer, W.R.
      Water sorption by corn starch and commercial modifications of
      starches.
      Industrial and Engineering Chemistry 36 (-) 316-319; 1944
      2,4,5
1741 Saltmarch, M.; Labuza, T.P.
      Influence of relative humidity (aw) on the physicochemical
      state of lactose in spray-dried sweet whey powders.
      Journal of Food Science 45 (5) 1231-1236, 1242; 1980
      4,5
1742 Salun, I.P.; Michailov, V.D.
      Equilibrium moisture content and coefficients of mass transfer
      of grains. (Orig. russ.)
      Pistschewaja Technol. 5 (-) 119-122; 1975
      5
1743 Salwin, H.
      Defining minimum moisture contents for dehydrated foods.
      Food Technology 13 (10) 594-595; 1959
      2,4
1744 Salwin, H.
      The role of moisture in deteriorative reactions of dehydrated
      foods.
      In: Fisher, F.R. (Ed.).
      Freeze Drying of Foods.
      National Academy of Sciences-National Research Council,
      Washington D.C.; 1962
      4,5
1745 Salwin, H.
      Moisture levels required for stability in dehydrated foods.
      Food Technology 17 (9) 34-43; 1963
      4,5
1746 Salwin, H.; Slawson, V.
      Moisture transfer in combinations of dehydrated foods.
      Food Technology 13 (12) 715-717; 1959
      1,2,5
1747 Samuelsson, E.G.
      Water activity.(Orig. dan.)
      Maelkeritidende 84 (29/30) 705-714, 716-718; 1971
      1,4

1748 Sanders, T.; Davis, N.; Diener, U.
Effect of carbon dioxide, temperature and relative humidity on
production of aflatoxin in peanuts.
Journal of the American Oil Chemists' Society 45 (10) 683-685;
1968
4

1749 Sangster, J.A.; Lenzi, F.
On the choice of methods for the prediction of the
water-activity and activity coefficient for multicomponent
aqueous solutions.
Canadian Journal of Chemical Engineering 52 (-) 392-396; 1974
3

1750 Sanjeevi, R.; Ramanathan, N.; Viswanathan, B.
Pore size distribution in collagen fiber using water vapor
adsorption studies.
Journal of Colloid and Interface Science 57 (2) 208-211; 1976
3

1751 Sanjeevi, R.; Ramanathan, N.; Viswanathan, B.
Pore size distribution and sorption-desorption hysteresis with
water.
Journal of Colloid and Interface Science 67 (3) 541-542; 1978
2,3

1752 Sanjeevi, R.; Viswanathan, B.
Mobility of water vapor adsorbed on proteins and polypeptides.
Journal of Colloid and Interface Science 82 (2) 572-573; 1981
2

1753 San Jose, C.; Asp, N.G.; Burvall, A.; Dahlqvist, A.; Logetko, V.P.
Water sorption in lactose hydrolyzed dry milk.
Journal of Dairy Science 60 (10) 1539-1543; 1977
5

1754 Saravacos, G.D.
Freeze-drying rates and water sorption of model food gels.
Food Technology 19 (4) 193-197; 1965
5

1755 Saravacos, G.D.
Effect of the drying method on the water sorption of
dehydrated apple and potato.
Journal of Food Science 32 (1) 81-84; 1967
5

1756 Saravacos, G.D.
Sorption and diffusion of water in dry soybeans.
Food Technology 23 (11) 1477-1479; 1969
2,5

1757 Saravacos, G.D.; Charm, S.E.
Dehydration of fruits and vegetables.
Food Technology 16 (1) 91-93; 1962
2,5

1758 Saravacos, G.D.; Stinchfield, R.M.
Effect of temperature and pressure on the sorption of water
vapor by freeze-dried food materials.
Journal of Food Science 30 (-) 779-786; 1965
2,5

1759 Sargent, G.P.
Computation of vapor pressure, dew-point and relative
humidity from dry and wet-bulb temperatures.
Meteorological Magazine 109 (-) 238-246; 1980
3

1760 Sarghat, F.
Studies on two instruments for measurement of water activity.
(Orig. fr.)
Lettre d'Informations, Viandes et Produits Carnes 6 (-) 5-6;
1980
3

1761 Savolainen, K.; Pyysalo, H.; Kallio, H.
The influence of water activity on the stability of
vulgaxanthin I.
Zeitschrift für Lebensmittel-Untersuchung und -Forschung
167 (4) 250-251; 1978
4

1762 Scatchard, G.; Hamer, W.; Wood, S.
Isotonic solutions.
I. The chemical potential of water in aqueous solutions of
sodium chloride, potassium chloride, sulfuric acid,
sucrose, urea and glycerol at 25 °C.
Journal of the American Chemical Society 60 (12) 3061-3070;
1938
3

1763 Schaich, K.M.; Karel, M.
Free radicals in lysozyme reacted with peroxidizing methyl
linoleate.
Journal of Food Science 40 (-) 456-459; 1975
4

1764 Scharnbeck, M.
Über die Feuchtigkeitsbestimmung in gefriergetrockneten
Lebensmitteln.
Die Nahrung 10 (4) 371-376; 1966
3,5

1765 Scharnbeck, M.; Dehne, E.
Sorptionsisothermen gefriergetrockneter Lebensmittel.
Bedeutung - Ermittlung - Darstellung.
Die Nahrung 10 (4) 297-303; 1967
1,5

1766 Schauss, H.
Problematik der Wassergehaltsbestimmung von Feststoffen.
Vergleichende Untersuchungen nach thermischen und chemischen
Meßmethoden.
Chemie-Ingenieur-Technik 36 (5) 469-479; 1964
1,3

1767 Schein, R.D.
Comments on the moisture requirements of fungus germination.
Phytopathology 54 (-) 1427; 1964
4

1768 Schelhorn, M. von
Untersuchungen über den Verderb wasserarmer Lebensmittel durc
osmophile Mikroorganismen.
I. Verderb von Lebensmitteln durch osmophile Hefen.
Zeitschrift für Lebensmittel-Untersuchung und -Forschung
91 (-) 117-124; 1950
4

1769 Schelhorn, M. von
Untersuchungen über den Verderb wasserarmer Lebensmittel durc
osmophile Mikroorganismen.
II. Grenzkonzentrationen für den osmophilen Schimmelpilz
Aspergillus glaucus in Abhängigkeit vom pH-Wert des
Substrats.
Zeitschrift für Lebensmittel-Untersuchung und -Forschung
91 (-) 338-342; 1950
4

1770 Schelhorn, M. von
Control of microorganisms causing spoilage in fruit and
vegetable products.
Advances in Food Research 3 (-) 429-482; 1951
4,5

1771 Schenk, D.
Das Verhalten von Hartgelatine-Kapseln in feuchter Atmosphäre.
Dissertation, Universität Marburg; 1974
2,3,5

1772 Scheuplein, R.J.; Morgan, L.J.
"Bound water" in keratin membranes measured by a microbalance
technique.
Nature 214 (-) 456-458; 1967
2,3

1773 Schierbaum, F.
Die Hydratation der Stärke.
I. Entwicklung und Stand der Kenntnisse über die Wasser-
Sorption und -Desorption der Stärke.
Die Stärke 12 (8) 237-243; 1960
5

1774 Schierbaum, F.
Die Hydratation der Stärke.
II. Tensimetrische Untersuchungen der Ad- und Desorption von
Wasser durch Stärke.
Die Stärke 12 (9) 257-265; 1960
3,5

1775 Schierbaum, F.; Täufel, K.
Die Hydratation der Stärke.
IV. Die Abhängigkeit der Sorptionswärme der Stärke von ihrer
Hydratation.
Die Stärke 14 (7) 233-238; 1962
2

1776 Schierbaum, F.; Täufel, K.
Zum Stand der Kenntnisse über die Wechselwirkung zwischen
nativer Stärke und Wasser.
Ernährungsforschung 7 (-) 647-679; 1963
2,5

1777 Schierbaum, F.; Ulmann, M.
Die Hydratation der Stärke.
III. Entwässerung der Stärke unter dem Einfluß von
Infrarot-Strahlen.
Die Stärke 14 (5) 161-167; 1962
2

1778 Schimpfky, S.
Das hygroskopische Verhalten ausgewählter
Futtermittelkomponenten.
Getreidewirtschaft (-) 225-228; 1976
1,5

1779 Schlünder, E.U.
Einfaches Verfahren zur Messung von Dampfdrücken.
In: Schlünder, E.U.
Über die Trocknung ruhender Einzeltropfen und fallender
Sprühnebel.
Dissertation, Technische Hochschule Darmstadt; 1962
3

1780 Schlünder, E.U.
A simple procedure for measurement of vapor pressure over
aqueous salt solutions.
In: Wexler, A.; Wildhack, W.A. (Edts.). Humidity and Moisture.
Vol. 3, Fundamentals and Standards. pp 535-544. Reinhold Publ.
Corporation, New York; 1965
3

1781 Schmidhofer, T.; Egli, H.R.
Zur Wasseraktivität von Fleischwaren.
Alimenta 5 (5) 169-170; 1972
4

1782   Schneeberger, R.; Voilley, A.; Weisser, H.
        Activity of water below 0 °C.
        In: Freezing, Frozen Storage and Freeze-Drying.
        International Institute of Refrigeration, 1977-1, pp 73-85.
        Paris; 1977
        1,2
1783   Schneeberger, R.; Voilley, A.; Weisser, H.
        Activity of water in frozen systems.
        International Journal of Refrigeration 1 (4) 201-206; 1978
        1,2
1784   Schneider, A.
        Untersuchungen über das charakteristische Trocknungsverhalten
        von Luzerne und Zuckerrübenblatt.
        Dissertation, Technische Hochschule München; 1954
        2,5
1785   Schneider, A.
        Neue Diagramme zur Bestimmung der relativen Luftfeuchtigkeit
        über gesättigten, wässrigen Salzlösungen und wässrigen
        Schwefelsäurelösungen bei verschiedenen Temperaturen.
        Holz als Roh- und Werkstoff 18 (-) 269-272; 1960
        3
1786   Schneider, A.S.
        Hydration of biological membranes.
        In: Rockland, L.B.; Stewart, G.F. (Edts.).
        Water Activity: Influences on Foodquality. pp 377-405.
        Academic Press, New York; 1981
        2,5
1787   Schneider, F.H.
        Zur Wasserbestimmung in vegetabilen Ölrohstoffen.
        Fette, Seifen, Anstrichmittel 83 (9) 329-337; 1981
        2,3
1788   Schnickels, R.A.; Warmbier, H.C.; Labuza, T.P.
        Effect of protein substitution on non-enzymatic browning in an
        intermediate moisture food system.
        Journal of Agricultural and Food Chemistry 24 (5) 901-903; 197
        4
1789   Schobert, B.
        The importance of water activity and water structure during
        hyperosmotic stress in algae and higher plants.
        Biochemie und Physiologie der Pflanzen 175 (-) 91-103; 1980
        1,2
1790   Schoebel, T.; Tannenbaum, S.; Labuza, T.P.
        Reaction at limited water concentration.
        1. Sucrose hydrolysis.
        Journal of Food Science 34 (4) 324-329; 1969
        4
1791   Schoeber, W.
        Regular regimes in sorption processes.
        Ph.D. Thesis, Eindhoven University of Technology; 1976
        1,2
1792   Schoof, H.F.
        The effects of various relative humidities on the life
        processes of the southern cow-pea weevil, Callosobruchus
        maculatus (Fabr.) at 30 +- 0.8 °C.
        Ecology 22 (-) 297-305; 1941
        4
1793   Schopper, E.
        Über ein direkt anzeigendes Feuchtigkeitsmeßgerät nach der
        Taupunktsmethode.
        Wissenschaftliche Arbeiten des Deutschen Meteorologischen
        Dienstes 1 (-) 138-143; 1947
        3

1794 Schornick, G.
Einfluß der Wasseraktivität auf Mikroorganismenwachstum und
Enzymaktivität.
In: Loncin, M. (Ed.).
Hochschulkurs Ausgewählte Themen der
Lebensmittelverfahrenstechnik "Wasseraktivität".
Institut für Lebensmittelverfahrenstechnik Universität
Karlsruhe; 1980
4

1795 Schram, A.
Theoretical considerations on the analysis of physical
adsorption isotherms.
Nuovo Cimento, Suppl. 5 (-) 309-320; 1967
2

1796 Schrenk, W.G.; Andrews, A.C.; King, H.H.
Heat of hydration of certain wheat flours and gluten.
Cereal Chemistry 26 (-) 51-59; 1949
2

1797 Schricker, J.A.
The respiration and storage behavior of flaxseed.
M.S. Thesis, University of Minnesota; 1948
4,5

1798 Schroeder, P. von
Über Erstarrungs- und Quellungserscheinungen von Gelatine.
Zeitschrift für Physikalische Chemie 45 (1) 75-117; 1903
2

1799 Schultes, H.
Wasseraufnahme von Kondensatorpapieren in Abhängigkeit von der
Temperatur und dem Wasserdampfdruck in der Umgebung.
Forschungsberichte des Wirtschafts- und Verkehrsministeriums
Nordrhein-Westfalen, Nr. 189, pp 19-23.
Westdeutscher Verlag, Köln und Opladen; 1955
3

1800 Schwarz, T.A.
Improvement needed in technic for testing food packages.
Food Industries 15 (-) 68-69, 124; 1943
4,5

1801 Schwieter, A.
Betrachtungen über Weißzuckertrockner und Kühler in Verbindung
mit Betonsilos.
Zeitschrift für die Zuckerindustrie 6 (10) 534-536; 1956
2,5

1802 Schwimmer, S.
Influence of water activity on enzyme reactivity and
stability.
Food Technology 34 (5) 64-74, 82-83; 1980
4

1803 Scott, J.L.
Preliminary observations on the moisture content and
hygroscopicity of cocoa beans.
Yearbook Dep.Agric. Gold Coast. (18 pp); 1928
5

1804 Scott, V.N.; Bernard, D.T.
Influence of temperature on the measurement of water activity
of food and salt systems.
Journal of Food Science 48 (2) 552-554; 1983
3

1805 Scott, W.J.
Water relations of Staphylococcus aureus at 30 °C.
Australian Journal of Biological Sciences 6 (-) 549-564; 1953
4

1806 Scott, W.J.
      Water relations of food spoilage microorganisms.
      Advances in Food Research 7 (-) 83-127; 1957
      1,4
1807 Scriban, R.
      Lagerung und Trocknung von Braugerste.
      Brauwissenschaft 18 (2) 41-48; 1965
      2,4,5
1808 Sebestyen, E.J.
      Aus Theorie und Praxis der Getreidetrocknung.
      Die Müllerei 41 (-) 558-559; 1958
                  43 (-) 600-601; 1958
      2,5
1809 Seehof, J.M.; Keilin, B.; Benson, S.W.
      The surface areas of proteins.
      V. The mechanisms of water sorption.
      Journal of the American Chemical Society 75 (-) 2427-2430;
      1953
      2,5
1810 Seidl, G.; Sczigel, R.
      Meß- und Auswertmethoden für die Bestimmung von
      Sorptionsisothermen.
      Luft- und Kältetechnik 4 (-) 114-117; 1966
      3
1811 Seiler, D.A.L.
      Equilibrium relative humidity of baked products with
      particular reference to the shelf life of cakes.
      In: Relative humidity in the food industry
      B.F.M.I.R.A.-Symposium Proc., No. 4, pp 28-36. Leatherhead,
      London; 1969
      4,5
1812 Seiler, D.A.L.
      The stability of intermediate moisture foods with respect to
      mould growth.
      In: Davies, R.; Birch, G.G.; Parker, K.J. (Edts.).
      Intermediate Moisture Foods. pp 166-181. Applied Science Publ.
      London; 1976
      4
1813 Seiler, D.A.L.
      Microbiological spoilage of bakery products.
      Swiss Food 3 (9a) 42-44; 1981
      4
1814 Senhaji, A.F.
      Protection des microorganismes par les matières grasses au
      cours des traitements thermiques.
      Ph.D. Thesis, University of Paris VI; 1973
      4
1815 Senhaji, A.F.
      The protective effect of fat on the heat resistance of
      bacteria. II.
      Journal of Food Technology 12 (3) 217-230; 1977
      4
1816 Senhaji, A.F.; Bimbenet, J.J.; Le Maguer, M.
      Protection des microorganismes par les matières grasses au
      cours des traitements thermiques. Partie 2.
      Industries Alimentaires et Agricoles 93 (1) 13-20; 1976
      4
1817 Senhaji, A.F.; Loncin, M.
      Protection des microorganismes par les matières grasses au
      cours des traitements thermiques. Partie 1.
      Industries Alimentaires et Agricoles 92 (6) 611-617; 1975
      4

1818 Senhaji, A.F.; Loncin, M.
      The protective effect of fat on the heat resistance of
      bacteria. I.
      Journal of Food Technology 12 (3) 203-216; 1977
      4
1819 Seow, C.C.
      Reactants mobility in relation to chemical reactivity in low-
      and intermediate moisture systems.
      Journal of the Science of Food and Agriculture 26 (4) 535-536;
      1975
      2,4
1820 Seow, C.C.; Goh, S.A.
      Lipid oxidation in the chinese sausage.
      Malaysian Agricultural Research 5 (2) 163-169; 1976
      4
1821 Seow, C.C.; Teng, T.T.
      The prediction of water activity of some supersaturated
      non-electrolyte aqueous binary solutions from ternary data.
      Journal of Food Technology 16 (6) 597-607; 1981
      3
1822 Sethi, R.K.; Chopra, S.L.
      Sorption of water vapor by wheat flours.
      Journal of Food Science and Technology India 10 (3) 9-12; 1973
      2,5
1823 Sethi, R.K.; Soni, G.L.; Chopra, S.L.
      Varietal differences in chickpea flour (Cicer arietinum) in
      relation to water vapor sorption.
      Cereal Chemistry 54 (1) 79-92; 1977
      2,5
1824 Seytre, G.; May, J.F.; Vallet, G.
      Etude de l'interaction eau-liaison peptide.
      I. -Etude de la sorption et de l'état de l'eau dans la
      polyglycine.
      Journal of Chemical Physics 69 (131) 958-963; 1972
      2,5
1825 Shanbhag, S.; Steinberg, M.P.; Nelson, A.I.
      Bound water defined and determined at constant temperature by
      wide-line NMR.
      Journal of Food Science 35 (5) 612-615; 1970
      2
1826 Shankman, S.; Gordon, A.R.
      The vapor pressure of aqueous solutions of sulfuric acid.
      Journal of the American Chemical Society 61 (-)2370-2373; 1939
      3
1827 Shapero, M.; Nelson, D.A.; Labuza, T.P.
      Ethanol inhibition of Staphylococcus aureus at limited water
      activity.
      Journal of Food Science 43 (5) 1467-1469; 1978
      4
1828 Sharp, J.G.
      Non-enzymic browning deterioration in dehydrated meat.
      In: Hawthorn, J.; Leitch, J.M. (Edts.).
      Recent Advances in Food Science. Vol. 2, pp 65-73.
      Butterworth, London; 1962
      4
1829 Sharp, P.F.; Doob, H. jr.
      Effect of humidity on moisture content and forms of lactose in
      dried whey.
      Journal of Dairy Science 24 (-) 679-690; 1941
      5

1830 Shaw, T.M.
The surface area of crystalline egg albumin.
The Journal of Chemical Physics 12 (9) 391-392; 1944
3

1831 Shaw, T.M.; Elsken, R.H.
Nuclear magnetic resonance absorption in hygroscopic
materials.
The Journal of Chemical Physics 8 (8) 1113-1114; 1950
2

1832 Shaw, T.M.; Elsken, R.H.
Investigation of proton magnetic resonance line width of
sorbed water.
The Journal of Chemical Physics 21 (3) 565-566; 1953
2

1833 Shaw, T.M.; Elsken, R.H.
Determination of water content of solids by nuclear magnetic
resonance.
American Chemical Society Abstracts of papers, 123 d meeting
March. pp 15-19; 1953
2,3

1834 Shaw, T.M.; Elsken R.H.
Techniques for nuclear magnetic resonance measurements on
granular hygroscopic materials.
Journal of Applied Physics 26 (3) 313-317; 1955
2,3

1835 Shaw, T.M.; Elsken, R.H.; Kunsman, C.H.
Moisture determination of foods by hydrogen nuclear magnetic
resonance.
Journal of the Association of Official Agricultural Chemists
36 (4) 1070-1076; 1953
2,3

1836 Shaw, T.M.; Vorkoeper, A.R.; Dyche, J.K.
Determination of surface area of dehydrated egg powder.
Food Research 11 (3) 187-194; 1946
3

1837 Shdanow, S.P.
Strukturuntersuchung poröser Adsorbenzien nach der
Kapillarkondensationstheorie.
In: Witzmann, H. (Ed.). Methoden der Strukturuntersuchung an
hochdispersen und porösen Stoffen. Akademieverlag, Berlin;
1961
2

1838 Sheffer, H.; Janis, A.A.; Ferguson, J.B.
The activity of water in sulphuric acid solutions at 25°C
by the isopiestic method.
Canadian Journal of Research 17 (-) 336-340; 1939
3

1839 Shelef, L.; Mohsenin, N.N.
Moisture relations in germ, endosperm and whole corn kernel.
Cereal Chemistry 43 (-) 347-353; 1966
5

1840 Sheppard, S.E.; Newsome, P.T.
The sorption of water by cellulose.
Industrial Engineering Chemistry 26 (-) 285-290; 1934
5

1841 Shiba, K.; Tozawa, T.
Wet-and-dry-plate dew-point hygrometer.
In: Wexler, A.; Ruskin, R.E. (Edts.).
Humidity and Moisture. Measurement and Control in Science and
Industry. Vol. 1, pp 64-75. Reinhold Publ. Corp., New York; 19
3

1842 Shibata, S.; Toyoshima, H.; Imai, T.; Inoue, Y.
Studies on storage of dried Japanese noodle.
Part I. Relation between NaCl content of dried noodle (udon)
and its equilibrium moisture.(Orig. jap.)
Nippon Shokuhin Kogyo Gakkaishi 23 (-) 397; 1976
5

1843 Shimazu, F.; Sterling, C.J.
Dehydration in model systems: cellulose and calcium pectinate.
Journal of Food Science 26 (-) 291-296; 1961
2,3

1844 Shishatskii, Y.I.; Kravchenko, V.M.
Isotherms of sorption of bakers' yeast.(Orig. russ.)
Khlebopekarnaya i Konditerskaya Promyshlennost' 13 (1) 25-27;
1969
5

1845 Shivashankar, S.; Govindajaran, V.S.
Equilibrium relative humidity relationships of processed
arecanut and whole dried ripe nuts.
Food Science, Mysore 12 (-) 317-321; 1963
3,5

1846 Shorter, S.A.; Hall, W.J.
The hygroscopic capacity of wool in different forms and its
dependence on atmospheric humidity and other factors.
Journal of the Textile Institute 15 (-) 305-327; 1924
2

1847 Shotton, E.; Harb, N.
The effect of humidity and temperature on the equilibrium
moisture content of powders.
Journal of Pharmacy and Pharmacology 17 (-) 504-508; 1965
5

1848 Shove, G.C.
Die Kühllagerung von Mais und die Kühltrocknung.
(Übersetzung aus: American Miller and Processor 95 (3) 28-30;
1967) Die Mühle 108 (33) 482-484; 1971
2,5

1849 Siddappa, G.S.; Nanjundaswamy, A.M.
Equilibrium relative humidity (ERH) relationships of fruit
juice and custard powders.
Food Technology 14 (10) 533-537; 1960
5

1850 Sidwell, C.G.; Salwin, H.; Koch, R.B.
The molecular oxygen content of dehydrated foods.
Journal of Food Science 27 (-) 255-261; 1962
5

1851 Signer, R.; Gál, S.
Untersuchungen der Bindung von Natrium- und Chlorionen durch
Casein mittels Wasserdampf-Sorptionsmessungen.
Die Makromolekulare Chemie 44-46 (-) 259-268; 1961
5

1852 Silver, M.E.
The behaviour of invertase in model systems at low moisture
contents.
Ph.D. Thesis, Massachusetts Institute of Technology; 1976
4

1853 Simatos, D.
Pression de vapeur saturante des produits alimentaires
congelés.
Revue Générale du Froid et des Industries Frigorifiques
62 (3) 183-185; 1971
3,5

1854 Simatos, D.; Blond, G.
The porous texture of freeze dried products.
In: Goldblith, S.A.; Rey, L.; Rothmayr, W.W. (Edts.).
Freeze Drying and Advanced Food Technology. pp 401-412.
Academic Press, London; 1975
4,5

1855 Simatos, D.; Faure, M.; Bonjour, E.; Couach, M.
The physical state of water at low temperatures in plasma with
different water contents as studied by differential thermal
analysis and differential scanning calorimetry.
Cryobilogy 12 (-) 202-208; 1975
2

1856 Simatos, D.; Faure, M.; Bonjour, E.; Couach, M.
Differential thermal analysis and differential scanning
calorimetry in the study of water in foods.
In: Duckworth, R.B. (Ed.).
Water Relations of Foods. pp 193-209.
Academic Press, London; 1975
2

1857 Simatos, D.; Le Meste, M.; Petroff, D.; Halphen, B.
Use of electron spin resonance for the study of solute
mobility in relation to moisture content in model food
systems.
In: Rockland, L.B.; Stewart, G.F. (Edts.).
Water Activity: Influences on Foodquality. pp 319-346.
Academic Press, New York; 1981
2,5

1858 Simonot-Grange, M.H.; Cointot, A.
Evolution des propriétés d'adsorption de l'eau par la
heulandite en relation avec la structure cristalline.
Bulletin de la Société Chimique, France (2) 421-427; 1969
2

1859 Simpson, W.T.
Predicting equilibrium moisture content of wood by
mathematical models.
Wood and Fiber 5 (1) 41-49; 1973
2

1860 Simpson, W.T.
Sorption theories applied to wood.
Wood and Fiber 12 (3) 183-195; 1980
2

1861 Sing, K.S.
The monolayer capacity in the physical adsorption of gases on
solids.
Chemistry and Industry, London 40 (-) 321-322; 1964
2

1862 Sing, K.S.W.
Empirical method for analysis of adsorption isotherms.
Chemistry and Industry 44 (11) 1520-1521; 1968
1,2

1863 Singh, B.P.N.; Narain, M.; Singh, H.
Kinetics of water vapor sorption by wheat flour from
saturated atmosphere.
Journal of Food Science and Technology 18 (5) 201-206; 1981
1

1864 Singh, R.K.; Lund, D.B.; Bülow, F.H.
Storage stability of intermediate apples: kinetics of quality
change.
Journal of Food Science 48 (3) 939-944; 1983
4

1865 Singh, R.S.; Ojha, T.P.
Equilibrium moisture content of groundnut and chillies.
Journal of the Science of Food and Agriculture 25 (5) 451-459;
1974
3,5
1866 Sinskey, A.J.
New developments in intermediate moisture foods: Humectants.
In: Davies, R.; Birch, G.G.; Parker, K.J. (Edts.).
Intermediate Moisture Foods. pp 260-280.
Applied Science Publ., London; 1976
4
1867 Skujins, J.J.; McLaren, A.D.
Enzyme reaction rates at limited water activities.
Science 158 (-) 1569-1570; 1967
4,5
1868 Skujins, J.; McLaren, A.D.
Urease reaction rates at low water activity.
Space Life Sciences 3 (-) 3-11; 1971
4
1869 Sloan, A.E.
Properties and effects of humectants in intermediate moisture
foods.
Ph.D. Thesis, University of Minnesota Minneapolis, USA; 1977
2,4,5
1870 Sloan, A.E.; Labuza, T.P.
Humectant water sorption isotherms.
Food Product Development 9 (12) 68; 1975
5
1871 Sloan, A.E.; Labuza, T.P.
Investigating alternative humectants for use in foods.
Food Product Development 9 (9) 75-88; 1975
4,5
1872 Sloan, A.E.; Labuza, T.P.
Prediction of water activity lowering ability of food
humectants at high aw.
Journal of Food Science 41 (3) 532-535; 1976
2,4,5
1873 Sloan, A.E.; Schlueter, D.; Labuza, T.P.
Effect of sequence and method of addition of humectants and
water on aw lowering ability in an IMF system.
Journal of Food Science 42 (1) 94-96; 1977
4
1874 Sloan, A.E.; Waletzko, P.T.; Labuza, T.P.
Effect of order-of-mixing on aw lowering ability of food
humectants.
Journal of Food Science 41 (3) 536-540; 1976
2,5
1875 Slyusarenko, S.N.
The hygroscopic properties of artificial sago.(Orig. russ.)
Sakharnaya Promyshlennost' 47 (2) 70-73; 1973
5
1876 Smejkalova, Z.; Hampl, J.
Sorption properties of cereals. I. General.(Orig. czech.)
Mlynsko-Pekarensky Prumysl 20 (12) 372-375; 1974
1,5
1877 Smejkalova, Z.; Hampl, J.
Sorption properties of cereals.
II. Measurement of adsorption-desorption isotherms of water
vapor at 20 °C.(Orig. czech.)
Mlynsko-Pekarensky Prumysl 21 (2) 38-42; 1975
1,5

1878   Smith, A.; Menzies, A.W.C.
       Studies in vapor pressure: II. A simple dynamic method,
       applicable to both solids and liquids for determining vapor
       pressures, and also boiling points at standard pressures.
       Journal of the American Chemical Society 32 (-) 907-914; 1910
       3
1879   Smith, A.K.; Nash, A.M.
       Water absorption of soybeans.
       Journal of the American Oil Chemists' Society 38 (-) 120-123;
       1961
       2
1880   Smith, D.S.; Mannheim, C.H.; Gilbert, S.G.
       Water sorption isotherms of sucrose and glucose by inverse gas
       chromatography.
       Journal of Food Science 46 (-) 1051-1053; 1981
       3,5
1881   Smith, F.B. jr.; Tsao, G.T.
       Sorption characteristics of a modified starch.
       Chemical Engineering Progr. Symp. Ser. 67 (-) 24-29; 1971
       2,3,5
1882   Smith, J.L.; Benedict, R.C.; Palumbo, S.A.
       Relationship of water activity to prevention of heat injury in
       Staphylococcus aureus.
       Lebensmittel-Wissenschaft und -Technologie 16 (4) 195-197; 1983
       4
1883   Smith, P.R.
       A new apparatus for the study of moisture sorption by starches
       and other foodstuffs in humidified atmospheres.
       In: Wexler, A.; Wildhack, W.A. (Edts.).
       Humidity and Moisture. Vol. 3, pp 487-494. Reinhold Publ.
       Corporation, New York; 1965
       3,5
1884   Smith, P.R.
       The determination of equilibrium relative humidity of water
       activity in foods - a literature review.
       Scientific and Technical Surveys, British Food Manufacturing
       Industries Research Association No. 70; 1971
       1,2,3
1885   Smith, S.E.
       The sorption of water vapor by high polymers.
       Journal of the American Chemical Society 69 (3) 646-651; 1947
       2,5
1886   Snow, D.
       Mould deterioration of feedingstuffs in relation to humidity
       of storage.
       III. The isolation of mould species from feedingstuffs stored
       at different humidities.
       Annales of Applied Biology 32 (-) 40; 1945
       4
1887   Snow, D.
       The germination of mould spores at controlled humidities.
       Annales of Applied Biology 36 (1) 1-13; 1949
       4
1888   Snow, D.; Crichton, M.H.G.; Wright, N.C.
       Mould deterioration of feeding-stuffs in relation to humidity
       of storage.
       Part I: The growth of moulds at low humidities.
       Annales of Applied Biology 31 (-) 102-116; 1944
       4,5

1889 Snyder, L.D.; Maxcy, R.B.
Effect of aw of meat products on growth of radiation resistant
Moraxella-Acinetobacter.
Journal of Food Science 44 (1) 33-36, 42; 1979
4

1890 Soekarto, S.T.
Water relations in food constituents and their application to
the development of a high protein, intermediate moisture,
soybean food.
Ph.D. Thesis, University of Illinois, Urbana, USA; 1978
1,2,4,5

1891 Soekarto, S.T.; Steinberg, M.P.
Determination of binding energy for the three fractions of
bound water.
In: Rockland, L.B.; Stewart, G.F. (Edts.).
Water Activity: Influences on Foodquality. pp 265-279.
Academic Press, New York; 1981
2,5

1892 Sörenfors, P.; Änäs, A.
Apparatus for measuring the humidity of air above 100 °C.
Laboratory Practice 26 (-) 549-550; 1977
3

1893 Soffer, A.; Folman, M.
Surface conductivity and conduction mechanisms on adsorption
of vapors on silica.
Transaction of the Faraday Society 62 (-) 3559-3569; 1966
2

1894 Sokolovsky, A.
Effect of humidity on hygroscopic properties of sugars and
caramel.
Industrial and Engineering Chemistry 29 (12) 1422-1423; 1937
2,5

1895 Sollars, W.F.
Water retention properties of wheat flour fractions.
Cereal Chemistry 50 (-) 717; 1973
2

1896 Solomon, M.E.
Experiments on the effects of temperature and humidity on the
survival of Halotydeus destructor (Tucker), Acarina fam.
Penthaleidae.
Australian Journal of Experimental Biology and Medical Science
15 (-) 1-16; 1937
4

1897 Solomon, M.E.
The use of cobalt salts as indicators of humidity and
moisture.
Annales of Applied Biology 32 (-) 75-85; 1945
3

1898 Solomon, M.E.
Control of humidity with potassium hydroxide, sulphuric acid,
or other solutions.
Bulletin of Entomological Research 42 (-) 543-554; 1951
3

1899 Solomon, M.E.
Estimation of humidity with cobalt thiocyanate papers and
permanent colour standards.
Bulletin of Entomological Research 48 (-) 489-506; 1957
3

1900 Somade, B.
Moisture equilibrium of palm kernels.
Journal of the Science of Food and Agriculture 6 (8) 425-427;
1955
5

1901 Sood, V.C.; Heldman, D.R.
Analysis of a vapor pressure manometer for measurement of
water activity in nonfat dry milk.
Journal of Food Science 39 (5) 1011-1013; 1974
3

1902 Sosedov, N.; Gospadinova, V.; Ryazantseva, M.
Kinetics of sorption of moisture by rice groats.(Orig. russ.)
Mukomol'no elevatornaya u Kombikormovaya Promyshlennost'
9 (-) 19-20; 1973
5

1903 Speakman, J.B.
An analysis of the water adsorption isotherm of wool.
Transaction of the Faraday Society 40 (-) 6-10; 1944
2,5

1904 Spencer, H.M.
Laboratory methods for maintaining constant humidity.
International Critical Tables 1 (-) 67; 1926
3

1905 Speranski, A.
Über den Dampfdruck und über die integrale Lösungswärme der
gesättigten Lösungen.
Zeitschrift für Physikalische Chemie 79 (-) 86; 1912
3

1906 Sperber, W.H.
Influence of water activity on food-borne bacteria - a review.
Journal of Food Protection 46 (2) 142-150; 1983
4

1907 Spicher, G.
Untersuchungen über das Vorkommen von Aflatoxin im Brot.
Zentralblatt für Bakteriologie, Abt. II 124 (-) 697-706; 1970
4

1908 Spieß, W.E.L.
Water activity in foodstuffs.
In: Johnson, A.H.; Peterson, M.S. (Edts.). Encyclopedia of
Food Technology and Food Science Series 2 (-) 948-954.
AVI Publ. Comp. Inc., Westport, Conn., USA; 1974
1

1909 Spieß, W.E.L.; Sole, P.; Pritzwald-Stegmann, B.F.
Wasserdampf-Sorptionsisothermen und spezifische Oberfläche
einiger wichtiger Lebensmittel.
Deutsche Lebensmittel-Rundschau 65 (4) 115-120; 1969
3,5

1910 Spieß, W.E.L.; Wolf, W.
Die Messung und Darstellung von
Wasserdampf-Sorptionsisothermen.
In: Forschungsberichte aus dem Gebiet der Lufttechnik.
pp 74-105.
Forschungsvereinigung für Luft- und Trocknungstechnik e.V.
Frankfurt/M; 1969
1,2,3,5

1911 Spieß, W.E.L.; Wolf, W.R.
The results of the COST 90 project on water activity.
In: Jowitt, R.; Escher, F.; Hallström, B.; Meffert, H.F.T.;
Spieß, W.E.L.; Vos, G. (Edts.).
Physical Properties of Foods. pp 65-87.
Applied Science Publ., London, New York; 1983
3,5

1912 Spinelli, J.; Groninger, H.; Koury, B.; Miller, R.
Functional protein isolates and derivates from fish muscle.
Process Biochemistry (12) 31-36, 42; 1975
4,5

1913 Spinelli, J.; Koury, B.
Nonenzymic formation of dimethylamine in dried fishery
products.
Journal of Agricultural and Food Chemistry 27 (5) 1104-1108;
1979
4

1914 Sponsler, O.L.; Bath, J.D.; Ellis, J.W.
Water bound to gelatin as shown by molecular structure
studies.
Journal of Physical Chemistry 44 (-) 996-1006; 1940
2

1915 Sposito, G.; Babcock, K.L.
Equilibrium theory of the kaolinite-water system at low
moisture contents, with some remarks concerning adsorption
hysteresis.
In: Bailey, S.W. (Ed.).
Clays and Clay Minerals 14 (-) 133-147, Pergamon Press, London;
1964 (Pub. 1966)
2,5

1916 Sprenger, J.
Einige Aspekte der Getreidetrocknungsanlagen unter besonderer
Berücksichtigung ihrer Verwendbarkeit für die verschiedenen
Arten landwirtschaftlicher Produkte.
Landwirtschaftliche Veröffentlichung der O.E.E.C., 27-34; 1953
5

1917 Stahl, P.H.
Trocknen von Tablettengranulaten mittels ablufttemperatur
geregelter Gleichgewichtseinstellung.
Chemie-Ingenieur-Technik 55 (3) 221; 1983
2,3

1918 Stamm, A.; Hansen, L.A.
Surface bound versus capillary condensed water in wood.
Journal of Physical Chemistry 42 (-) 209-214; 1938
2

1919 Steinbach, G.
Zur Bindung des Wassers in trocknenden Substanzen.
Vortragsmanuskript Arbeitssitzung des GVC(Gesellschaft für
Verfahrenstechnik und Chemieingeneurwesen,Düsseldorf)
-Fachausschusses Trocknungstechnik Bad Hersfeld; 1976
2

1920 Steinberg, M.P.; Leung, H.
Some applications, of wide-line and pulsed NMR in
investigations of water in foods.
In: Duckworth, R.B. (Ed.).
Water Relations of Foods. pp 233-248.
Academic Press, London; 1975
2

1921 Sterling, C.; Masuzawa, M.
Gel/water relationships in hydrophilic polymers: Nuclear
magnetic resonance.
Die Makromolekulare Chemie 116 (-) 140; 1963
2

1922 Stevens, N.E.
A method for studying the humidity relations of fungi in
culture.
Phytopathology 6 (-) 428-432; 1916
4

1923 Stille, B.
Der mikrobielle Verderb getrockneter Lebensmittel in
Abhängigkeit von der relativen Luftfeuchtigkeit.
Vorratspflege und Lebensmittelforschung 5 (-) 403-408; 1942
4

1924 Stille, B.
      Grenzwerte der relativen Luftfeuchtigkeit und des
      Wassergehaltes getrockneter Lebensmittel für den mikrobiellen
      Befall.
      Zeitschrift für Lebensmittel-Untersuchung und -Forschung
      88 (-) 9-14; 1948
      4,5
1925 Stille, B.; Uzelac, G.
      Das Verhalten von Mikroorganismen in Abhängigkeit von der
      Wasseraktivität.
      Zeitschrift für Pflanzenkrankheiten und Pflanzenschutz
      85 (34) 186-190; 1978
      4
1926 Stitt, F.
      Moisture equilibrium and the determination of water content of
      dehydrated foods.
      In: Fundamental aspects of the dehydration of foodstuffs.
      pp 67-88. Conference in Aberdeen , march 1958,
      Society of Chemical Industry, London; 1958
      2,3,5
1927 Stokes, R.H.; Robinson, R.A.
      Standard solutions for humidity control at 25 °C.
      Industrial and Engineering Chemistry 41 (9) 2013; 1949
      3
1928 Stokes, R.H.; Robinson, R.A.
      Interactions in aqueous nonelectrolyte solutions.
      I. Solute-solvent equilibria.
      Journal of Physical Chemistry 70 (7) 2126-2131; 1966
      3
1929 Stoloff, L.
      Calibration of water activity measuring instruments and
      devices: collaborative study.
      Journal of Association of Official Analytical Chemists
      61 (5) 1166-1178; 1978
      3,5
1930 Storey, R.M.; Stainsby, G.
      The equilibrium water vapor pressure of frozen cod.
      Journal of Food Technology 5 (2) 157-163; 1970
      5
1931 Strasser, J.
      Detection of quality changes in freeze-dried beef by
      measurement of the sorption isobar hysteresis.
      Journal of Food Science 34 (1) 18-21; 1969
      2,3,5
1932 Streit, K.; Rüegg, M.; Blanc, B.
      Beeinflussung des Wachstums von Milchsäure- und Propionsäure-
      Bakterien durch die Wasseraktivität in Abhängigkeit des
      Zusatzes zum Nährmedium.
      Milchwissenschaft 34 (8) 459-462; 1979
      4
1933 Strohman, R.D.
      Thermodynamic properties of water in corn.
      M.S. Thesis, University of Illinois, Urbana, USA; 1965
      2,5
1934 Strohman, R.D.; Yoerger, R.R.
      A new equilibrium moisture-content equation.
      Transaction of the American Society of Agricultural Engineers
      10 (5) 675-677; 1967
      2,5

1935 Strolle, E.O.; Cording, J. jr.
Moisture equilibria of dehydrated potato flakes.
Food Technology 19 (5) 171-173; 1965
5

1936 Strolle, E.O.; Cording, J. jr.; McDowell, P.E.; Eskew, R.K.
Effect of sucrose on crispness of explosion puffed apple
pieces exposed to high humidities.
Journal of Food Science 35 (4) 338-342; 1970
5

1937 Strong, D.H.; Foster, E.F.; Duncan, C.C.
Influence of water activity on the growth of Clostridium
perfringens.
Applied Microbiology 19 (6) 980-987; 1970
4

1938 Stute, R.
Wechselwirkungen zwischen Wasser und Kohlenhydraten.
Lebensmittel-Technologie 13 (1) 3-13; 1980
2,4

1939 Suarez, C.; Aguerre, R.; Viollaz, P.E.
Analysis of the desorption isotherms of rough rice.
Lebensmittel-Wissenschaft und -Technologie 16 (3) 176-179;
1983
2,5

1940 Suarez, C.; Viollaz, P.; Chirife, J.
Diffusional analysis of air drying of grain sorghum.
Journal of Food Technology 15 (5) 523-531; 1980
2

1941 Subasi, A.R.
The water activity (aw) requirements of xerotolerant molds
isolated from cured and aged meat products.
Thesis, Iowa State University, Ames, USA; 1965
4,5

1942 Suggett, A.
Water carbohydrate interactions.
In: Duckworth, R.B. (Ed.).
Water Relations of Foods. pp 23-36.
Academic Press, London; 1975
2,4

1943 Suggett, A.
The significance of water, hydration and aqueous systems in
intermediate moisture foods.
In: Davies, K.; Birch, G.G.; Parker, K.J. (Edts.).
Intermediate Moisture Foods. pp 66-74.
Applied Science Publ., London; 1976
4

1944 Supplee, G.C.
Humidity equilibria of milk powders.
Journal of Dairy Science 9 (-) 50-61; 1926
5

1945 Susi, H.; Ard, J.S.; Carroll, R.J.
The infrared spectrum and water binding of collagen as a
function of relative humidity.
Biopolymers 10 (9) 1597-1604; 1971
2

1946 Suzuki, T.
State of water in sea food.
In: Rockland, L.B.; Stewart, G.F. (Edts.).
Water Activity: Influences on Foodquality. pp 743-763.
Academic Press, New York; 1981
2

1947 Swan, E.; Urquhart, A.R.
      Adsorption equations - A review of the literature.
      Journal of Physical Chemistry 31 (-) 251; 1927
      2
1948 Szalai, L.
      Apparatus for the assessment of equilibrium relative humidity
      in the temperature range of 0 to -30 °C.
      In: Wexler, A.; Ruskin, R.E. (Edts.).
      Humidity and Moisture. Measurement and Control in Science and
      Industry. Vol. 1, pp 298-301. Reinhold Publ. Corporation,
      New York; 1965
      3
1949 Szulmayer, W.
      Humidity and moisture measurement.
      Food Research Quarterly 29 (2) 27-35; 1969
      3
1950 Szulmayer, W.
      Water in the vapor phase and its interaction with foods.
      Food Research Quarterly 33 (1) 19-25; 1973
      2,5
1951 Tabouret, T.
      Rôle de l'activité de l'eau dans la cristallisation du miel.
      Apidologie 10 (4) 341-358; 1979
      4
1952 Tändler, K.; Rödel, W.
      Herstellung und Haltbarkeit von dünnkalibrigen Dauerwürsten.
      II. Haltbarkeit.
      Die Fleischwirtschaft 63 (2) 150-162; 1983
      4
1953 Takahashi, K.; Shirai, K.; Wada, K.; Kawamura, A.
      Thermal behaviour of high molecular weight substances in
      foods.
      IV. Thermal behaviour of proteins at relatively low moisture
      content. (Orig. jap.)
      Journal of the Agricultural Chemical Society of Japan (Nihon
      Nogei Kagakkai-shi) 54 (5) 357-359; 1980
      4,5
1954 Takizawa, A.
      Adsorption of water by proteins.(Orig. jap.)
      Kobunshi 181 (16) 499-506; 1967
      5
1955 Tamburini, U.
      Sorption isotherms and swelling of wood impregnated with
      P.E.G. (polyethylene glycol).
      Ricerca Scientifica 38 (12) 1194-1198; 1968
      2,5
1956 Tamsma, A.; Pallansch, M.J.
      Factors related to the storage stability of foam-dried whole
      milk.
      IV. Effect of powder moisture content and in-pack oxygen at
      different storage temperatures.
      Journal of Dairy Science 47 (-) 970; 1964
      4
1957 Tarasenkov, D.N.
      Vapor pressure of aqueous sulfuric acid solutions.
      Journal of Applied Chemistry (USSR) 28 (-) 1053-1058; 1955
      3
1958 Tarasov, S.G.
      Hygroscopic properties and thermodynamic characteristics of
      moisture mass exchange of maize starch.(Orig. russ.)
      Pistschewaja Technol. - (5) 116-121; 1972
      2,5

1959 Taylor, A.A.
Determination of moisture equilibria in dehydrated foods.
Food Technology 15 (12) 536-540; 1961
2,3,5

1960 Taylor, A.; Rowlinson, J.
The thermodynamic properties of aqueous solutions of glucose.
Transaction of the Faraday Society 51 (-) 1183-1192; 1955
2,3

1961 Taylor, N.W.; Cluskey, J.E.; Senti, F.R.
Water sorption by dextrans and wheat starch at high humidities.
Journal of Physical Chemistry 65 (10) 1810-1816; 1961
5

1962 Taylor, N.W.; Zobel, H.F.; Hellman, N.N.; Senti, F.R.
Effects of structure and crystallinity on water sorption of
dextrans.
Journal of Physical Chemistry 63 (-) 599-603; 1959
4,5

1963 Teixeira Neto, R.O.; Denizo, N.; Quast, D.G.
Water activity of intermediate moisture foods.
Coletanea do Instituto de Tecnologia de Alimentos
7 (1) 191-207; 1976
4,5

1964 Teixeira Neto, R.O.; Quast, D.G.
Isotermas de adsorcao de unidade en dimentos.
Coletanea do Instituto de Tecnologia de Alimentos 8 (1)
141-197; 1977
5

1965 Teng, T.T.; Lenzi, F.
Water activity data representation of aqueous solutions
at 25 °C.
Canadian Journal of Chemical Engineering 52 (-) 387-391; 1974
3

1966 Teng, T.T.; Seow, C.C.
A comparative study of methods for prediction of water
activity of multicomponent aqueous solutions.
Journal of Food Technology 16 (4) 409-419; 1981
2

1967 Ter-Minassian-Saraga, L.
Protein denaturation on adsorption and water activity at
interfaces: An analysis and suggestion.
Journal of Colloid and Interface Science 80 (2) 393-401; 1981
2

1968 Ternstroem, A.; Nickels, C.
Effect of water activity on the heat resistance of bacteria.
(Orig. swed.)
Livsmedelsteknik 17 (6) 245-248; 1975
4

1969 Tessem, B.M.; Hughes, F.J.
Description and test evaluation of the Honeywell relative
humidity flour moisture meter.
Cereal Science Today 10 (2) 50-52; 1965
3

1970 Texter, J.A.; Kellerman, R.; Klier, K.
Water sorption in dextran gel.
Carbohydrate Research 41 (-) 191-210; 1975
5

1971 Thakker, M.T.; Chi, C.W.; Pech, R.E.; Wasan, D.T.
Vapor pressure measurements of hygroscopic salts.
Journal of Chemical and Engineerings Data 13 (-) 533; 1968
3

1972 Thalmann, A.; Wolf, W.
Das Wasserdampfsorptionsverhalten verschiedener
Handelsfuttermittel pflanzlicher und tierischer Herkunft.
Landwirtschaftliche Forschung 33 (4) 349-360; 1980
3,5

1973 Theimer, O.
Hüttigs adsorption isotherm.
Nature (London) 168 (-) 873; 1951
2

1974 Thieme, E.
Die Bestimmungen des GF-Wertes und sein Einfluß auf die
Haltbarkeit des Produktes.
Gordian 72 (4) 132-135; 1972
3,5

1975 Thieme, J.C.
The hygroscopic properties of raw sugars and molasses.
International Sugar Journal 36 (-) 192-193; 1934
5

1976 Thijssen, H.A.C.
The influence of processing on food quality.
In: Proc. 6th European Symposium - Food. pp 9-36; 1975
4

1977 Thijssen, H.A.C.
Optimization of process conditions during drying with regard
to quality factors.
Lebensmittel-Wissenschaft und - Technologie 12 (6) 308-317;
1979
2,5

1978 Thijssen, H.A.C.; Kerkhof, P.J.A.M.
Effect of temperature and water concentration during
processing on food quality.
Journal of Food Process Engineering 1 (2) 129-147; 1977
4

1979 Thomas, J.M.; Williams, B.R.
Application of a metal vacuum microbalance to the study of
solid surfaces by physical adsorption.
In: Waters, P.M. (Ed.).
Vacuum Microbalance Techniques. Vol. 4, p. 209.
Plenum Press, New York; 1965
3

1980 Thompson, H.J.; Shedd, C.K.
Equilibrium moisture and heat of vaporization of shelled corn
and wheat.
Agricultural Engineering, London 35 (-) 786-788; 1954
5

1981 Thung, S.;
Moisture determination in dried vegetables.
Journal of the Science of Food and Agriculture 15 (4) 237-244;
1964
3

1982 Tietke, H.W.
Studien zu den Möglichkeiten der Senkung von Warenverlusten an
feuchteempfindlichen Lebensmitteln unter besonderer
Berücksichtigung des Sorptionsverhaltens hygroskopischer
Lebensmittel und des notwendigen Verpackungseinsatzes.
Dissertation, Karl-Marx-Universität Leipzig; 1968
4,5

1983 Tietke, H.W.
Sorptionsverhalten hygroskopischer Lebensmittel und
notwendiger Verpackungseinsatz.
Die Verpackung 10 (1) 10-12; 1969
4

1984 Tietke, H.W.
Lagerung von Trockenfrüchten in Einzelhandelsverpackungen.
Lebensmittel-Industrie 17 (11) 419-421; 1970
4,5

1985 Tietke, H.W.
Beobachtungen zur Wanderung von Feuchtefronten in lose
lagernden schüttbaren Lebensmitteln.
Lebensmittel-Industrie 17 (12) 454-456; 1970
2,5

1986 Tilbury, R.H.
The microbial stability of intermediate moisture foods with
respect to yeasts.
In: Davies, R.; Birch, G.G.; Parker, K.J. (Edts.).
Intermediate Moisture Foods. pp 138-165.
Applied Science Publ., London; 1976
4

1987 Tilenschi, S.
Kelvin relation and discontinuities of the adsorption
isotherm.(Orig. rom.)
Analele Universitatii Bucuresti Seria Stiintele Naturii Chimie
17 (1) 13-20; 1968
2

1988 Timbers, G.E.
Measurement of moisture in foods.
Instrumentation in the Food and Beverage Industry 2 (-) 45-49;
1973
3

1989 Tjaberg, T.B.
Measurement of the water activity of foods.(Orig. norw.)
Norsk Veterinaertidsskrift 87 (2) 95-104; 1975
3,4

1990 To, E.C.; Flink, J.M.
"Collapse", a structural transition in freeze dried
carbohydrates.
II. Effect of solute composition.
Journal of Food Technology 13 (6) 567-581; 1978
4

1991 Todes, O.M.; Lezin, Y.S.
Dynamics of sorption and desorption in a fluidized bed, taking
into consideration the heat evolved.
Zurnal Prikladnoj Chimii 40 (-) 2280-2285; 1967
1,2

1992 Tokariev, P.
Equilibrium moisture content of sunflower varieties with high
oil content.
Mukomol' no-Elevatornaya Kombikormovaya Promyslennost
1 (-) 28-29; 1973
In: Gough, M.C.; Lipiatt, G.A. (Edts.).
Moisture Humidity Equilibria of Tropical Stored Produce.
Part II - Oilseeds.
Tropical Stored Products Information 34, 49-61; 1977
5

1993 Toledo, R.T.
Determination of water activity in foods.
In: Proc. The Meat Industry Research Conference.
University of Chicago.
American Meat Institute Foundation. pp 85-106; 1973
3

1994 Toledo, R.; Steinberg, M.P.; Nelson, A.
Quantitative determination of bound water by NMR.
Journal of Food Science 33 (3) 315-317; 1968
2

1995 Tomas, B.
Stabilization of foods by reducing the water activity.
Technologija Mesa 13 (7/8) 203-207; 1972
4

1996 Tomás, J.; Pinaga, F.; Lafuente, B.; Primo, E.
Liofilizacion de triturados integrales de citricos.
I. Estudios sobre su comportamiento eutectico e higroscópico.
Revista de Agroquimica y Technologia de Alimentos
9 (3) 406-414; 1969
2,5

1997 Tome, D.; Bizot, H.; Guilbot, A.; Multon, J.L.; Drapron, R.
Les aliments à humidité intermédiaire.
Série Synthéses Bibliographiques No. 16, APRIA, CDIUPA,
Massy, France; 1978
1,2,4,5

1998 Tome, D.; Nicolas, J.; Drapron, R.
Influence of water activity on the reaction catalyzed by
polyphenoloxidase (E.C. 1.14.181) from mushrooms in organic
liquid media.
Lebensmittel-Wissenschaft und -Technologie 11 (1) 38-41; 1978
4

1999 Tomka, I.
Untersuchungen über die Wasserdampf-Sorption quellbarer
Körper.
Inaugural Dissertation, Universität Bern; 1973
1,2,3

2000 Tomkins, R.G.; Mapson, L.W; Allen, R.J.L.; Wager, H.G.;
Barker, J.
The drying of vegetables.
III. The storage of dried vegetables.
Journal of the Society of Chemical Industry; 63 (-) 225-231;
1944
4

2001 Toupin, C.J.; Le Maguer, M.; McGregor, J.R.
The evaluation of BET-constants from sorption isotherms data.
Lebensmittel-Wissenschaft und -Technologie 16 (3) 153-156;
1983
2

2002 Traktvenko, A.I.
Kinetics of water vapor sorption by dried wheat grain.
Tr. VNII Zerna i Produktov ego Pererab. 97 (-) 6-11; 1981
2

2003 Tran, T.T.; Lenzi, F.
Methods of estimating the water activity of supersaturated
aqueous solutions.
Canadian Journal of Chemical Engineering 52 (-) 798-802; 1974
3

2004 Travaglini, D.; Tosello, Y.
Aplicacao da equacao de Henderson em estudos de umidade de
equilibrio em café em coco, despolpado e beneficiado.
Coletanea do Instituto de Tecnologia de Alimentos
2 (-) 403-413; 1967/1968
2,5

2005 Trenk, H.L.; Hartman, P.A.
Effects of moisture content and temperature on aflatoxin
production in corn.
Applied Microbiology 19 (-) 781-784; 1970
4

2006 Troesch, A.
Contribution à une généralisation des equations d'adsorption
multimoléculaire de Brunauer, Emmett et Teller et de Hüttig.
Journal de Chimie Physique 48 (9-10) 454-464; 1951
2
2007 Troller, J.A.
Effect of water activity on enterotoxin B production and
growth of Staphylococcus aureus.
Applied Microbiology 21 (-) 435-439; 1971
4
2008 Troller, J.A.
Effect of water activity on enterotoxin A production and
growth of Staphylococcus aureus.
Applied Microbiology 24 (-) 440-443; 1972
4
2009 Troller, J.A.
Effect of water activity and pH on staphylococcal enterotoxin
B production.
Acta Alimentaria Academiae Scientiarum Hungaricae
2 (-) 351-360; 1973
4
2010 Troller, J.A.
The water relations of food-borne bacterial pathogens.
A review.
Journal of Milk and Food Technology 36 (5) 276-288; 1973
4
2011 Troller, J.A.
Statistical analysis of aw measurements obtained with the Sina
scope.
Journal of Food Science 42 (1) 86-90; 1977
3
2012 Troller, J.A.
Food spoilage by microorganisms tolerating low - aw
environments.
Food Technology 33 (1) 72-75; 1979
4
2013 Troller, J.A.
Evaluation of a modified Sina/Beckman hygrometer.
Journal of Food Protection 42 (3) 208-210; 1979
4
2014 Troller, J.A.
Influence of water activity on microorganisms in foods.
Food Technology 34 (5) 76-80, 82; 1980
4
2015 Troller, J.A.
Methods to measure water activity.
Journal of Food Protection 46 (2) 129-134; 1983
3
2016 Troller, J.A.
Water activity measurements with a capacitance manometer.
Journal of Food Science 48 (3) 739-741; 1983
3
2017 Troller, J.A.; Christian, J.H.B.
Water Activity and Food.
Academic Press, London; 1978
1,2,3,4
2018 Troller, J.A.; Stinson, J.V.
Influence of water activity on growth and enterotoxin
formation by Staphylococcus aureus in foods.
Journal of Food Science 40 (4) 802-804; 1975
4

2019 Tsai, W.Y.J.; Moy, J.H.; Nip, W.K.; Frank, H.A.
Aspergillus parasiticus growth and aflatoxin production on
dehydrated taro.
Journal of Food Science 46 (-) 1167-1169; 1981
4,5

2020 Tschapek, M.; Wasowski, C.
Determination of bound and free water in electrolyte solutions
by surface tension measurements.
Journal of Colloid and Interface Science 60 (1) 205-206; 1977
2

2021 Tsutsumi, C.; Koizumi, H.; Tani, T.
Hygroscopic equilibrium and water absorption characteristics
of rough rice.(Orig. jap.)
Report of the Food Research Institute (Shokuryo Kenkyusho
Kenkyu Hokoku) 25 (-) 44-49; 1970
5

2022 Tuerlinckx, G.; Berckmans, D.; Goedseels, V.
Desorption isotherms of malt.
Brauwissenschaft 35 (3) 70-74; 1982
3,5

2023 Tuite, J.; Foster, G.H.
Effect of artificial drying on the hygroscopic properties of
corn.
Cereal Chemistry 40 (6) 630-637; 1963
2

2024 Ubertis, B.; Roversi, G.
The water sorption of corn starch.
Die Stärke 5 (10) 266-267; 1953
5

2025 Uboldi Eiroa, M.N.
Atividade de agua: influencia sobre o desenvolvimento de
microorganismos e metodos de determinacao em alimentos.
Boletim ITAL, Campinas 18 (3) 353-383; 1981
1,4

2026 Udani, K.H.; Nelson, A.I.; Steinberg, M.P.
Rate of moisture adsorption by wheat flour and its relation to
physical, chemical and baking characteristics.
Food Technology 22 (12) 65-68; 1968
2,4,5

2027 Urone, P.; Takahashi, Y.; Kennedy, G.H.
Sorption isotherms on liquid-coated adsorbents.
Analytical Chemistry 40 (-) 1130-1134; 1968
3

2028 Urquhart, A.R.
Adsorption hysteresis.
Journal of Textile Institute (British) 20 (1) T 117; 1929
2

2029 Urquhart, A.R.; Eckersall, N.
The adsorption of water by rayon.
Journal of Textile Institute Transactions T 163-T174; 1932
5

2030 Urquhart, A.R.; Williams, A.M.
The moisture relations of cotton.The effect of temperature on
the absorption of water by soda-boiled cotton.
Journal of the Textile Jnstitute 15, T559 -T572; 1924
2,3,5

2031 Uzelac, G.
Die Überlebensfähigkeit von Bakterien fäkalen Ursprungs in
Trockenprodukten in Abhängigkeit von der Wasseraktivität.
3. Mitteilung: Überlebensfähigkeit in Milchpulver.
Deutsche Lebensmittel-Rundschau 74 (6) 228-229; 1978
4

2032 Uzelac, G.; Groneuer, K.J.
Die Überlebensfähigkeit von Bakterien fäkalen Ursprungs in
Trockenprodukten in Abhängigkeit von der Wasseraktivität.
1. Mitteilung: Einleitung und Problemstellung.
Deutsche Lebensmittel-Rundschau 73 (8) 235-239; 1977
1,4
2033 Uzelac, G.; Stille, B.
Die Überlebensfähigkeit von Bakterien fäkalen Ursprungs in
Trockenprodukten in Abhängigkeit von der Wasseraktivität.
II. Eigene Versuche mit Escherichia coli und Streptococcus
faecalis.
Deutsche Lebensmittel-Rundschau 73 (10) 325-329; 1977
4
2034 Vaamonde, G.; Chirife, J.; Scorza, O.C.
An examination of the minimal water activity for
Staphylococcus aureus ATCC 6538 P growth in laboratory media
adjusted with less conventional solutes.
Journal of Food Sciences 47 (4) 1259-1262; 1982
4
2035 Vaidya, P.S.; Verma, K.K.; Rustagi, K.N.; Jaisoni, J.C.;
Mathew, T.V.
Studies on the equilibrium relative humidity and seasonal
variaton in moisture content of walnuts (Juglans regia).
Journal of Food Science and Technology, India 14 (4) 169-173;
1977
5
2036 Vail, G.E.; Bailey, C.H.
The state of water in colloidal gels: free and bound water
in breaddoughs.
Cereal Chemistry 17 (-) 397-917; 1940
2
2037 Väisälä, V.
Mixing hygrostat for calibration of hygroscopic hygrometers.
In: Wexler, A.; Wildhack, W.A. (Edts.).
Humidity and Moisture. Vol. 3, pp 473-477. Reinhold Publ.
Corporation, New York; 1965
3
2038 Valentine, L.
Absolute crystallinities of celluloses from moisture sorption
determinations.
Chemistry and Industry 43 (3) 1279-1280; 1956
5
2039 Valentine, L.J.
Studies on the sorption of moisture by polymers.
I. Effect of crystallinity.
Journal of Polymer Science 27 (-) 313-333; 1958
2,5
2040 Van Aken, J.G.T.
The analysis of adsorption isotherms of vapors on homogeneous
surfaces.
Chemisches Laboratorium RVO-TNO, Report No. 1968-11
2
2041 Van den Berg, I.C.
Thermodynamical aspects of water activity in intermediate
moisture foods.
Course on Intermediate Moisture Foods CPCIA Europe,
Seminaire E. 5; 1975
1,4

2042 Van den Berg, I.C.
Water activity with special reference to food systems.
Paper presented at Fortbildungskurs Schweizerische
Gesellschaft für Lebensmittelwissenschaft und - Technologie.
Zürich; 1977
1,2

2043 Van den Berg, C.
Vapour sorption equilibria and other water - starch
interactions; a physico-chemical approach.
Thesis, Agricultural University Wageningen, The
Netherlands; 1981
1,2,5

2044 Van den Berg, C.
Description of water activity of foods for engineering purpos
by means of the G.A.B. model of sorption.
Paper presented at 3. Intern. Congr. on Engineering and Food.
Dublin; 1983
2

2045 Van den Berg, C.; Bruin, S.
Water activity and its estimation in food systems: Theoretica
aspects.
In: Rockland, L.B.; Stewart, G.F. (Edts.).
Water Activity: Influences on Foodquality. pp 1-61.
Academic Press, New York; 1981
1,2

2046 Van den Berg, C.; Kaper, F.S.; Weldring, J.A.G.; Wolters, I.
Water binding by potato starch.
Journal of Food Technology 10 (6) 589-602; 1975
2,3,5

2047 Van den Berg, C.; Leniger, H.A.
The water activity of foods.
In: Miscellaneous Paper 15, 231-244; 1978
Levensmiddelentechnologie van de Landbouwhogeschool,
Wageningen; 1978
1,2,5

2048 Van den Berg, L.; Lentz, C.P.
Effect of relative humidity, temperature and length of stora
on decay and quality of potatoes and onions.
Journal of Food Science 38 (-) 81-83; 1973
4

2049 Van der Held, E.F.M.
Symposion "Trocknen in der Lebensmittelindustrie".
II. Über die für das Trocknen wichtigen physikalischen
Eigenschaften von Lebensmitteln.
De Ingenieur 69 (6) 15-19; 1957
1,2

2050 Van Krevelen, D.W.; Hoftijzer, P.J.
Properties of polymers.
2nd rev. ed. Elsevier Scientific Publ. Co., Amsterdam, Oxfor
New York; 1976
5

2051 Vansteenkiste, J.P.; Hoof, J. van
A simplified method for the determination of water activity
(aw).
Proc. European Meeting of Meat Research Workers
24 (-) G20:1-G20:5; 1978
3

2052 Vansteenkiste, J.P.; Hoof, J. van
The course and significance of water activity during ripenin
of dry sausages.
Archiv für Lebensmittelhygiene 30 (4) 117-120; 1979
4

2053 Vansteenkiste, J.P.; Hoof, J. van
Measurement of water activity of dry sausages. A comparison of
the isopiestic method and the electric hygrometer.
Archiv für Lebensmittelhygiene 33 (5) 116-118; 1982
3,5

2054 Van Twisk, P.
The sorption isotherms of maize meal.
Journal of Food Technology 4 (1) 75-82; 1969
3,5

2055 Varshney, N.N.; Ojha, T.P.
Water vapor sorption properties of dried milk baby foods.
Journal of Dairy Research 44 (1) 93-101; 1977
2,3,5

2056 Veasey, D.P.
Effect of ambient conditions on oven methods of moisture
measurement.
Milling (London) 153 (1) 17-20; 1971
3

2057 Veerraju, P.
A new graphical method to predict the moisture pick-up or loss
by packaged foodstuffs.
Journal of Food Science and Technology (Mysore)
7 (Supplem.) 40-42; 1970
2,4

2058 Veerraju, P.; Hemavathy, J.; Prabhakar, J.V.
Influence of water activity on pellicle chaffing, color and
breakage of walnut (Juglans regia) kernels.
Journal of Food Processing and Preservation 2 (1) 21-31; 1978
4,5

2059 Verrips, C.T.; Kwast, R.H.
Heat resistance of Citrobacter freundii in media with various
water activities.
European Journal of Microbiology 4 (3) 225-231; 1977
4

2060 Vigo, M.S.; Chirife, J.; Scorza, O.C.; Cattaneo, P.; Bertoni,
M.H.; Sarrailh, P.
Estudio sobre alimentos tradicionales de humedad intermedia
elaborados en la Argentina.
I. Determinacion de actividad acuosa (aw), pH, humedad y
solidos solubles.
Revista de Agroquimica y Tecnologia de Alimentos 21 (1) 91-99;
1981
3,4

2061 Villota, R.; Saguy, I.; Karel, M.
An equation correlating shelf life of dehydrated vegetable
products with storage conditions.
Journal of Food Science 45 (-) 398-399, 401; 1980
4

2062 Vincent, S.F.; Bristol, K.E.
Equilibrium humidity measurement.
Industrial and Engineering Chemistry; Analytical Edition
17 (7) 465-466; 1945
3,5

2063 Viollaz, P.; Chirife, J.; Iglesias, H.A.
Slopes of moisture sorption isotherms of foods as a function
of moisture content.
Journal of Food Science 43 (2) 606-608; 1978
2

2064 Viollaz, P.E.; Suarez, C.; Alzamora, S.
Temperature prediction in air drying of food materials: a
simple method.
Journal of Food Technology 15 (4) 361-367; 1980
2

2065   Vogt, H.F.; Hammerschmidt, W.B.; Herrero, F.M.
       Verpackung von Trockensuppen.
       Zeitschrift für Lebensmittel-Technologie und Verfahrenstechni▮
       29 (3) 82-85; 1978
       3,4,5
2066   Voirol, F.A.
       Xylitol, its properties and applications.
       In: Birch, G.G.; Parker, K.J. (Edts.).
       Sugar: Science and Technology. pp 325-343.
       Applied Science Publ., London; 1979
       5
2067   Volgunov, G.
       Action of trypsin and the proteolytic pea enzyme on a dried
       substrate in various stages of humidity.
       Biokhimija 13 (-) 104-108; 1948
       4
2068   Volker, H.H.
       Equilibrium humidity as the factor determining the shelf life
       of confectionery goods.
       Food Technology of Australia 22 (2) 58-61; 1970
       1,4
2069   Volkov, K.P.; Savostianov, E.O.
       Vapor pressure of saturated sucrose solutions at low
       temperatures.(Orig. russ.)
       Univ. etat de Kiev, Bull. Sci. rec. chim. 3 (-) 103-119; 1937
       3,5
2070   Volkov, M.A.; Tserevitinov, O.B.; Mikhailov, V.D.
       Equilibrium moisture and thermodynamic parameters of
       granulated sugar.(Orig. russ.)
       Sakharnaya Promyshlennost' 47 (6) 43-45; 1973
       2,5
2071   Vollhardt, D.; Kretschmar, G.
       Untersuchungen zur Desorptionskinetik grenzflächenaktiver
       Stoffe.
       Abhandlungen der Deutschen Akademie der Wissenschaften. Klass▮
       für Chemie, Geologie und Biologie (6) 841-848; 1966
       2
2072   Vollmer, W.
       Der Transport von Gasen und Dämpfen in Papier.
       Chemie-Ingenieur-Technik 26 (2) 90-94; 1954
       2,5
2073   Volman, D.H.; Simons, J.W.; Seed, J.R.; Sterling, C.
       Sorption of water vapor by starch. Thermodynamics and
       structural changes for dextrin, amylose, and amylopectin.
       Journal of Polymer Science 46 (-) 355-364; 1960
       2,5
2074   Vos, P.T.; Labuza, T.P.
       Technique for measurement of water activity in the high aw
       range.
       Journal of Agricultural and Food Chemistry 22 (2) 326-327;
       1974
       3,5
2075   Vrchlabsky, J.; Leistner, L.
       Hitzeresistenz der Enterokokken bei unterschiedlichen
       aw-Werten.
       Die Fleischwirtschaft 50 (9) 1237-1238; 1970
       4
2076   Vrchlabsky, J.L.; Leistner, L.
       Beziehung zwischen Wasseraktivität und Wasserbindung von Rin▮
       und Schweinefleisch.
       Die Fleischwirtschaft 50 (7) 967-968; 1970
       2,5

2077 Vrchlabsky, J.; Leistner, L.
Erprobung eines SINA-Gerätes mit eingebautem Schreiber zur
Messung der Wasseraktivität von Fleisch und Fleischwaren.
Die Fleischwirtschaft 50 (8) 1106-1107; 1970
3,5

2078 Vrchlabsky, J.; Leistner, L.
Hitzeresistenz von Laktobazillen bei unterschiedlichen
aw-Werten.
Die Fleischwirtschaft 51 (9) 1368-1370; 1971
4

2079 Wahba, M.; Nashed, S.; Aziz, K.
Moisture relations of cellulose.
IV - The effect of stabilization of cotton cellulose on its
sorptivity at different temperatures and its bearing on
isosteric heats of sorption.
The Journal of the Textile Institute, Transactions
49 (11) T519-T531; 1958
2,3,5

2080 Wakamatsu, T.; Sato, Y.
Determination of unfreezable water in sucrose, sodium chloride
and protein solutions by differential scanning calorimeter.
Journal of the Agricultural Chemical Society of Japan
53 (12) 415-420; 1979
2

2081 Waldham, D.G.; Halvorson, H.O.
Studies on the realtionship between equilibrium vapor pressure
and moisture content of bacterial endospores.
Applied Microbiology 2 (-) 333-338; 1954
4

2082 Waletzko, P.; Labuza, T.P.
Accelerated shelf-life testing of an intermediate moisture
food in air and in oxygen-free atmosphere.
Journal of Food Science 41 (6) 1338-1344; 1976
4

2083 Walker, B.; Renken, A.
Direkte mikrogravimetrische Messung der Sorptionskinetik unter
Reaktionsbedingungen.
Chemie-Ingenieur-Technik 55 (10) 806-807; 1983
2,3

2084 Walker, J.E.; Wolf, M.; Kapsalis, J.G.
Adsorption of water vapor on myosin A and myosin B.
Journal of Agricultural and Food Chemistry 21 (5) 878-880;
1973
2,5

2085 Wallingford, L.; Labuza, T.P.
Evaluation of the water binding properties of food
hydrocolloids by physical/chemical methods and in a low fat
meat emulsion.
Journal of Food Science 48 (1) 1-5; 1983
2,5

2086 Warburton, S.; Pixton, S.W.
The effect of the addition of glycerol on the moisture
content/equilibrium relative humidity relationship of
wheatfeed.
Journal of Stored Products Research 11 (-) 107-109; 1975
5

2087 Warburton, S.; Pixton, S.W.
Moisture relations of dried onions.
Journal of Stored Products Research 13 (2) 85-86; 1977
4,5

2088 Warburton, S.; Pixton, S.W.
The moisture relations of spray dried skimmed milk.
Journal of Stored Products Research 14 (2/3) 143-158; 1978
5

2089 Warburton, S.; Pixton, S.W.
The significance of moisture in dried milk.
Dairy Industries International 43 (4) 23, 26-27; 1978
5

Warburton, S.   see also 2200

2090 Ward, C.B. jr.; Tischer, R.G.
Use of cobaltous chloride to detect moisture patterns in
partially dehydrated kernels of corn.
Cereal Chemistry 30 (-) 420-426; 1953
3

2091 Warmbier, H.
Non enzymic browning kinetics of an intermediate moisture
model food system.
Ph.D. Thesis, University of Minnesota, USA; 1975
4

2092 Warmbier, H.C.; Schnickels, R.A.; Labuza, T.P.
Effect of glycerol on non-enzymatic browning in a solid
intermediate moisture model food system.
Journal of Food Science 41 (3) 528-531; 1976
4

2093 Warmbier, H.; Schnickels, R.; Labuza, T.P.
Non-enzymatic browning kinetics in an intermediate moisture
model system; effect of glucose to lysine ratio.
Journal of Food Science 41 (5) 981-983; 1976
4

2094 Warner, D.T.
Theoretical studies of water in carbohydrates and proteins.
In: Rockland, L.B.; Stewart, G.F. (Edts.).
Water Activity: Influences on Foodquality. pp 435-465.
Academic Press, New York; 1981
2

2095 Warner Cervone, N.; Harper, J.M.
Viscosity of an intermediate moisture dough.
Journal of Food Processing Engineering 2 (-) 83-95; 1978
4

2096 Waters, P.M.
Vacuum microbalance techniques. Vol. 4
Plenum Press, New York; 1965
3

2097 Watt, I.C.
Sorption of water vapor by keratin.
Journal of Macromolecular Science - Review
Macromolecular Chemistry C 18 (2) 169-245; 1980
1,2,5

2098 Watt, I.C.
The theory of water sorption by biological materials.
In: Jowitt, R.; Escher, F.; Hallström, B.; Meffert, H.F.T.;
Spieß, W.E.L.; Vos, G. (Edts.).
Physical Properties of Foods. pp 27-40.
Applied Science Publ., London, New York; 1983
2,5

2099   Weaver, B.H.; Riley, R.
        Measurement of water in gases by electrical conduction in a
        film of hygroscopic material and the use of pressure change
        in calibration.
        Journal of Research of the National Bureau of Standards
        40 (3) 169-214; 1948
        3
2100   Webb, S.J.
        Factors affecting the viability of air-borne bacteria.
        3. The role of bound water and protein structure in the death
        of air-borne cells.
        Canadian Journal of Microbiology 6 (-) 89-105; 1960
        2,4
2101   Webster, C.E.M.; Wood, R.M.; Ledward, D.A.
        Prediction of water activity aw in cook-soak equilibrated
        intermediate moisture meats.
        Meat Science 3 (1) 43-51; 1979
        2,4
2102   Wedler, G.
        Adsorption.
        Verlag Chemie, Weinheim; 1970
        1
2103   Weisser, H.
        Vliv teploty na aktivitu vody v potravinach.
        (Einfluß der Temperatur auf die Wasseraktivität von
        Lebensmitteln)
        Potravinarska a chladici Technika 9 (6) 620-625; 1978
        1,3,5
2104   Weisser, H.
        NMR-techniques in studying bound water in foods.
        In: Linko, P.; Mälkki, Y.; Olkku, J.; Larinkari, J. (Edts.).
        Food Process Engineering. Vol. 1, Food Processing Systems.
        pp 326-336. Applied Science Publ., London; 1980
        2
2105   Weisser, H.
        Die Technik der Sorptions- und Wasseraktivitätsmessung.
        In: Loncin, M. (Ed.).
        Hochschulkurs Ausgewählte Themen der
        Lebensmittelverfahrenstechnik "Wasseraktivität".
        Institut für Lebensmittelverfahrenstechnik Universität
        Karlsruhe; 1980
        3
2106   Weisser, H.; Bürkle, R.; Loncin, M.
        Messen von Sorptionsisothermen bei höheren Temperaturen.
        Zeitschrift für Lebensmittel-Technologie und
        -Verfahrenstechnik 29 (8) 310-314; 1978
        1,3,5
2107   Weisser, H.; Roth, T.; Plett, E.A.
        Die Bedeutung der Wasseraktivität in der
        Lebensmittelverfahrenstechnik. (Orig. czech.)
        Premysl Potravin 32 (11) 639-646; 1981
        1
2108   Weisser, H.; Weber, J.; Loncin, M.
        Wasserdampf-Sorptionsisothermen von Zuckeraustauschstoffen im
        Temperaturbereich von 25 bis 80 °C.
        Zeitschrift für Lebensmittel-Technologie und
        -Verfahrenstechnik 33 (2) 89-97; 1982
        3,5
2109   Weldring, J.A.G.
        Determination of the water activity of foods by dew point
        measurement.(Orig. dutch.)
        Voedingsmiddelentechnologie 10 (37) 13-17; 1978
        3

2110 Weldring, J.A.G.
      Determination of the water activity of foods on the basis of
      dew point measurements.
      II. Results for a range of products at various temperatures
      and moisture contents.(Orig. dutch.)
      Voedingsmiddelentechnologie 11 (21/22) 28-31; 1978
      2,5

2111 Weldring, J.A.G.; Wolters, J.; Van den Berg, C.
      Apparatus for the accurate determination of the isotherms for
      water vapor sorption in foods.
      Mededelingen Landbouwhogeschool Wageningen,75-18; 1975
      3

2112 Weronski, E.B.
      Adsorption and concomitant phenomena.
      I. Theoretical principles.
      Electrochimica Acta 14 (3) 231-240; 1969
      2

2113 Weston, W.J.; Morris, H.J.
      Hygroscopic equilibria of dry beans.
      Food Technology 8 (-) 353-355; 1954
      4,5

2114 Wexler, A.
      Calibration of humidity measuring instruments at the National
      Bureau of Standards.
      ISA Transactions 7 (4) 356; 1968
      3

2115 Wexler, A.
      Vapor pressure formulation for water in the range 0 to 100 °C
      A revision.
      Journal of Research of the National Bureau of Standards.
      Phys. Chem. 80 A (5/6) 775-785; 1976
      3

2116 Wexler, A.
      Vapor pressure formulation for ice.
      Journal of Research of the National Bureau of Standards. Phys
      Chem. 81 A (1) 5-20; 1977
      3

2117 Wexler, A.; Brombacher, W.G.
      Methods of measuring and testing hygrometers.
      National Bureau of Standards; Circ. 512-513; 1951
      3

2118 Wexler, A.; Greenspan, L.
      Vapor pressure equation for water in the range 0 to 100 °C.
      Journal of Research of the National Bureau of Standards (U.S.
      Phys. and Chem. 75A (3) 213-230; 1971
      3

2119 Wexler, A.; Hasegawa, S.
      Relative humidity - temperature relationships of some
      saturated salt solutions in the temperature range 0 to 50 °C.
      Journal of Research of the National Bureau of Standards
      53 (1) 19-26; 1954
      3

2120 Wexler, A.; Ruskin, R.
      Humidity and Moisture. Measurement and Control in Science and
      Industry. Vol. 1, Principles and Methods of Measuring Humidit
      in Gases.
      Reinhold Publ. Corporation, New York; 1965
      3

2121 Wheelock, T.D.; Lancaster, E.B.
      Thermal properties of wheat flour.
      Die Stärke 22 (2) 44-48; 1970
      2

2122 Whitbread, E.I.; Gastler, G.F.
      Hygroscopic moisture of grain sorghum and wheat as influenced
      by temperature and humidity.
      Proc. South Dakota Academy of Sciences 26 (-) 80-84; 1946/7
      5
2123 White, G.M.; Ross, I.J.; Klaiber, J.D.
      Moisture equilibrium in mixing of shelled corn.
      Transactions of the American Society of Agricultural Engineers
      15 (3) 508-509, 514; 1972
      2
2124 White, G.W.; Cakebread, S.H.
      The glassy state in certain sugar-containing food products.
      Journal of Food Technology 1 (1) 73-82; 1966
      4
2125 White, H.J.; Eyring, H.
      The adsorption of water by swelling high polymeric materials.
      Textile Research Journal 17 (10) 523; 1947
      2
2126 White, R.K.
      Swelling stress in corn kernel as influenced by moisture
      sorption.
      M.S. Thesis, Pennsylvania State University State Park, USA;
      1966
      2,4
2127 White, R.T.
      Studies on the storage and shipment of whole black pepper
      grown in the Orient.
      Journal of Economic Entomology 50 (4) 423-428; 1957
      4,5
2128 Whittier, E.O.; Gould, S.P.
      Vapor pressures of saturated equilibrated solutions of lactose,
      sucrose, glucose, galactose.
      The Journal of Industrial and Engineering Chemistry
      22 (1) 77-78; 1930
      3,5
2129 Widicus, W.A.; Kirk, J.R.; Gregory, J.F.
      Storage stability of alpha-tocopherol in a dehydrated model
      food system containing no fat.
      Journal of Food Science 45 (-) 1015-1018; 1980
      4
2130 Wiebe, H.H.
      Measuring water potential (activity) from free water to oven
      dryness.
      Plant Physiology 68 (6) 1218-1221; 1981
      3
2131 Wiebe, H.H.; Kidambi, R.N.; Richardson, G.H.; Ernstrom, C.A.
      A rapid psychrometric procedure for water activity measurement
      of foods in the intermediate moisture range.
      Journal of Food Protection 44 (12) 892-895; 1981
      3
2132 Wilbaux, R.; Hahn, D.
      Contribution à l'étude des phénomènes intervenant au cours de
      la conservation du café vert.
      Revue Café, Cacao, Thé, 10 (4) 342-367; 1966
      4,5
2133 Williams, C.J.
      A new lease on life for gravimetric adsorption.
      American Laboratory (6) 40-45; 1969
      3

2134 Williams, J.C.
Chemical and non-enzymic changes in intermediate moisture foods.
In: Davies, R.; Birch, G.G.; Parker, K.J. (Edts.).
Intermediate Moisture Foods. pp 100-119.
Applied Science Publ., London; 1976
4

2135 Williams, J.L.; Stannett, V.T.
Highly water absorptive cellulose by postdecrystallization.
Journal of Applied Polymer Science 23 (-) 1265-1268; 1979
5

2136 Willigen, A.H.A. de; Groot, P.W. de
Der Einfluß der Temperatur auf die Adsorption von Wasser und von wasserlöslichen Stoffen an unverkleisterter Kartoffelstärke.
Die Stärke 19 (11) 368-372; 1967
4

2137 Wilson, R.E.
Humidity control by means of sulfuric acid solutions, with critical compilation of vapor pressure data.
The Journal of Industrial and Engineering Chemistry
13 (4) 326-331; 1921
2,3

2138 Wilson, R.E.
Some new methods for the determination of the vapor pressure of salt-hydrates.
Journal of the American Chemical Society 43 (-) 704; 1921
3

2139 Wilson, R.E.; Fuwa, T.
Humidity equilibria of various common substances.
The Journal of Industrial and Engineering Chemistry
14 (10) 913-918; 1922
3,5

2140 Windisch, S.; Neumann-Duscha, I.
Hefen als Verderbniserreger von Süsswaren, unter Berücksichtigung osmophiler Hefen.
Alimenta 13 (2) 60-62; 1974
4

2141 Windle, J.J.
Sorption of water by wool.
Journal of Polymer Science 21 (7) 103-112; 1956
2,5

2142 Windle, J.J.; Shaw, T.M.
Dielectric properties of wool-water systems at 3000 and 9300 Megacycles.
Journal of Chemical Physics 22 (10) 1752-1757; 1954
3

2143 Windle, J.J.; Shaw, T.M.
Dielectric properties of wool-water systems.
II. 26000 Megacycles.
Journal of Chemical Physics 25 (3) 435-439; 1956
3

2144 Windsor, W.E.; Sobel, F.; Morris, V.B. jr.; Hooper, M.V.
Critical relative humidities of some salts.
Review of Scientific Instruments 24 (-) 334; 1953
3

2145 Wink, W.A.
Determining the moisture equilibrium curves of hygroscopic materials.
Industrial and Engineering Chemistry, Analytical Edition
18 (4) 251-252; 1946
3

2146 Wink, W.A.
      Moisture equilibrium: Sorption isotherms offer graphic method
      of finding relationship between moisture level and equilibrium
      relative humidity.
      Modern Packaging 20 (6) 135-138, 162, 164; 1947
      1
2147 Wink, W.A.; Bobb, F.C.; Van den Akker, J.A.
      Equilibrium moisture content determination at high
      temperature.
      Technical Association of the Pulp and Paper Industry
      41 (11) 643-646; 1958
      3
2148 Wink, W.A.; Sears, G.R.
      Instrumentation studies LVII. Equilibrium relative humidities
      above saturated salt solutions at various temperatures.
      Technical Association of the Pulp and Paper Industry
      33 (9) 96A-99A; 1950
      3
2149 Wink, W.A.; Van den Akker, J.A.
      A thin film psychrometer for measuring relative humidity in
      small spaces.
      Technical Association of the Pulp and Paper Industry
      39 (9) 647-649; 1956
      3
2150 Winkler, C.A.; Geddes, W.F.
      Heat of hydration of wheat flour and certain starches
      including wheat, rice and potato.
      Cereal Chemistry 8 (-) 455-475; 1931
      2
2151 Winston, P.W.; Bates, D.H.
      Saturated solutions for the control of humidity in biological
      research.
      Ecology 41 (1) 232-236; 1960
      3
2152 Woessner, D.E.; Snowden, B.S. jr.
      Pulsed NMR study of water in agar gels.
      Journal of Colloid and Interface Science 34 (2) 290-299; 1970
      2
2153 Woessner, D.E.; Snowden, B.S. jr.; Chiun, Y.C.
      Pulsed NMR study of the temperature hysteresis in the
      agar-water system.
      Journal of Colloid and Interface Science 34 (2) 283-289; 1970
      2
2154 Wojciechowski, J.
      Water activity.
      I. Significance of water activity in the production of meat
      products.(Orig. pol.)
      Gospodarka Miesna 30 (3) 20-22; 1978
      1,4
2155 Wojciechowski, J.
      Water activity.
      II. Non-instrumental methods for determination of water
      activity.(Orig. pol.)
      Gospodarka Miesna 30 (4) 25-27; 1978
      3,5
2156 Wolf, M.; Walker, J.E.; Kapsalis, J.G.
      Water vapor sorption hysteresis in dehydrated food.
      Journal of Agricultural and Food Chemistry 20 (5) 1073-1077;
      1972
      2,3,5

2157 Wolf, W.; Spieß, W.E.L.
Use of reference materials for the measurement of water
sorption isotherms.
In: Schmitt, B.F. (Ed.).
Production and Use of Reference Materials.
Proc. International Symposium 1979. pp 263-270.
Bundesanstalt für Materialprüfung, Berlin; 1980
3,5

2158 Wolf, W.; Spieß, W.E.L.; Jung, G.
Wasserdampf-Sorptionsisothermen von Lebensmitteln.
Berichtsheft 18 der Fachgemeinschaft Allgemeine Lufttechnik im
VDMA Frankfurt/Main; 1973
5

2159 Wolf, W.; Spieß, W.E.L.; Jung, G.
Die Wasserdampfsorptionsisothermen einiger in der Literatur
bislang wenig berücksichtigter Lebensmittel.
Lebensmittel-Wissenschaft und -Technologie 6 (3) 94-96; 1973
5

2160 Wolf, W.; Spieß, W.E.L.; Jung, G.; Weisser, H.; Bizot, H.;
Duckworth, R.B.
The water-vapor sorption isotherms of microcrystalline
cellulose (MCC) and of purified potato starch. Results of a
collaborative study.
Journal of Food Engineering 3 (1) 51-73; 1984
3,5

2161 Wolf, W.; Spieß, W.E.L.; Weisser, H.; Gál, S.; Bizot, H.
Mikrokristalline Zellulose als Referenzmaterial zum Bestimmen
des Wasserdampf-Sorptionsverhaltens von Lebensmitteln.
Zeitschrift für Lebensmittel-Technologie und Verfahrenstechnik
31 (3) 148-154; 1980
3,5

2162 Wootton, A.E.
Sorption isotherms of macadamia nuts (title incomplete)
East African Industrial Research Organization Annual Report,
6-9; 1968-69
In: Gough, M.C.; Lippiatt, G.A. (Edts.).
Moisture Humidity Equilibria of Tropical Stored Produce
II - Oilseeds.
Tropical Stored Products Information 34 (-) 49-61; 1977
5

2163 Wootton, A.E.
Coffee processing research.
1. East African Industrial Research Organization, Annual
Report, 33-34; 1968-69
In: Gough, M.C.; Lippiatt, G.A. (Edts.).
Moisture Humidity Equilibria of Tropical Stored Produce.
Part III - Legumes, Spices and Beverages.
Tropical Stored Products Information 35 (-) 15-29; 1978
5

2164 Wootton, M.; Reaoch, R.R.; Wrigley, C.W.; Gras, P.W.
Factors affecting the rehydration of dried gluten.
Food Technology in Australia 34 (4) 154-156; 1982
4

2165 Worral, R.W.
Psychrometric determination of relative humidities in air with
dry bulb temperatures exceeding 212 °F.
In: Wexler, A. Ruskin, R.E. (Edts.).
Humidity and Moisture. Measurement and Control in Science and
Industry. Vol. 1, pp 105-109. Reinhold Publ. Corporation,
New York; 1965
3

2166 Worthington, J.T.
Moisture content and texture of dried chili peppers stored in
polyethylene-lined burlap bags.
United States Department of Agriculture, Agricultural
Marketing Service, Report No. 444 p. 6; 1961
In: Gough, M.C.; Lippiatt, G.A. (Edts.).
Moisture Humidity Equilibria of Tropical Stored Produce.
Part III - Legumes, Spices and Beverages.
Tropical Stored Products Information 35 (-) 15-29; 1978
5

2167 Wright, M.E.; Porterfield, J.G.
Specific heat of Spanish peanuts.
Transactions of the American Society of Agricultural Engineers
13 (4) 508-510; 1970
2

2168 Wylie, R.G.
The properties of water salt systems in relation to
hygrometry.
In: Wexler, A.; Wildhack, W.A. (Edts.).
Humidity and Moisture. Vol. III, pp 507-517.
Reinhold Publ. Corporation, New York; 1965
3

2169 Yamada, J.; Niimura, Y.; Takasawa, K.
Precision of automatic water activity (aw) measuring device.
Ministry of Health and Welfare, Japan; (brochure,
published by Rotronic AG, Zürich, Switzerland); 1982
3

2170 Yano, T.; Kojima, I.; Torikata, Y.
Role of water in withering of leafy vegetables.
In: Rockland, L.B.; Stewart, G.F. (Edts.).
Water Activity: Influences on Foodquality. pp 765-790.
Academic Press, New York; 1981
2,4

2171 Yoshii, H.
Fermented foods and water activity.(Orig. jap.)
Journal of the Society of Brewing, Japan (Nihon Jozo Kyokai
Zasshi) 74 (4) 213-218; 1979
4

2172 Young, D.M.; Crowell, A.D.
Physical adsorption of gases.
Butterworth, London; 1962
1,2

2173 Young, J.F.
Humidity control in the laboratory using salt solutions.
A review.
Journal of Applied Chemistry 17 (9) 241-245; 1967
3

2174 Young, J.H.
A study of the sorption and desorption equilibrium moisture
content isotherms of biological materials.
Ph.D. Thesis, Oklahoma State University, USA; 1966
1,2,3,5

2175 Young, J.H.
Evaluation of models to describe sorption and desorption
equilibrium moisture content isotherm of Virginia-type
peanuts.
Transactions of the American Society of Agricultural Engineers
19 (1) 146-150,155; 1976
2,5

2176   Young, J.H.; Nelson, G.L.
       Theory of hysteresis between sorption and desorption isotherms
       in biological materials.
       Transactions of the American Society of Agricultural Engineers
       10 (2) 260-263; 1967
       2,5
2177   Young, J.H.; Nelson, G.L.
       Research on hysteresis between sorption and desorption
       isotherms of wheat.
       Transactions of the American Society of Agricultural Engineers
       10 (6) 756-761; 1967
       2,5
2178   Youngquist, G.R.; Allen, J.L.; Eisenberg, J.
       Adsorption of hydrocarbons by synthetic zeolites.
       Industrial and Engineering Chemistry Product Research and
       Development 10 (3) 308-314; 1971
       3
2179   Zabik, M.E.; Fierke, S.G.; Bristol, D.K.
       Humidity effects on textural characteristics of sugar snap
       cookies.
       Cereal Chemistry 56 (1) 29-33; 1979
       4,5
2180   Zama, K.; Takama, K.; Mizushima, Y.
       Effects of metal salts and antioxidants on the oxidation of
       fish lipids during storage under the conditions of low and
       intermediate moistures.
       International Congress of Food Science & Technology, Kyoto,
       Japan Abstracts, p. 202; 1978
       4
2181   Zelery, L.
       Moisture measurements in the grain industry.
       Cereal Science Today 5 (-) 130-136; 1960
       3
2182   Zettlemoyer, A.C.
       Hydrophobic surfaces.
       Journal of Colloid and Interface Science 28 (3/4) 343-369;
       1968
       2
2183   Zettlemoyer, A.; Micale, F.; Klier, K.
       Adsorption of water on well characterized solid surfaces.
       In: Franks, F. (Ed.).
       Water. A Comprehensive Treatise. Vol. 5. Water in disperse
       systems.
       Plenum Press, New York; 1975
       2
2184   Zettlemoyer, A.C.; Iyengar, R.D.; Scheidt, P.
       Heats of immersion and water sorption studies on bare and
       silica-coated rutile surfaces.
       Journal of Colloid and Interface Science 22 (-) 172-178; 1966
       2
2185   Zhushman, A.I.; Syroedov, V.I.; Kovalenok, V.A.
       Thermodynamic characteristics of moisture transfer in various
       starches.(Orig. russ.)
       Sakharnaya Promyshlennost' 1 (-) 62-64; 1977
       2
2186   Zimm, B.H.; Lundberg, J.L.
       Sorption of vapors by high polymers.
       Journal of Physical Chemistry 60 (4) 425; 1956
       2

2187 Zimmerli, A.
Quantitative Bestimmung von wenigen osmotoleranten Hefen in Lebensmitteln.
Alimenta 19 (-) 67-71; 1980
4

2188 Zitzmann, W.
Darstellung, Annäherung und Auswertung von Sorptionsisothermen.
In: Loncin, M. (Ed.).
Hochschulkurs Ausgewählte Themen der Lebensmittelverfahrenstechnik "Wasseraktivität".
Institut für Lebensmittelverfahrenstechnik Universität Karlsruhe; 1980
2

2189 Zorbalas, D.I.
Bestimmung der Bindungswärme des Wassers im Tabak.
Beiträge zur Tabakforschung 4 (7) 301-307; 1968
2,5

2190 Zschaler, R.
Nachweis von Mikroorganismen.
Untersuchungstechniken und Beurteilungskriterien bei getrockneten sowie bei begasten Lebensmitteln.
Die Ernährungswirtschaft 26 (9) 61-67; 1979
4

2191 Zsigmondy, R.
Über die Struktur des Gels der Kieselsäure. Theorie der Entwässerung.
Zeitschrift für Anorganische und Allgemeine Chemie 71 (-) 356-377; 1911
2

2192 Zürcher, K.; Hadorn, H.
Wasserbestimmung in Lebensmitteln nach der Methode von Karl Fischer.
Deutsche Lebensmittel-Rundschau 74 (8) 287-296; 1978
3

2193 Zürcher, K.; Hadorn, H.
Ausgleichsfeuchtigkeit und Gehalt von ätherischem Öl von Gewürzen des Handels.
Deutsche Lebensmittel-Rundschau 77 (7) 239-245; 1981
4,5

2194 Zürcher, K.; Hadorn, H.
Wasserbestimmung nach Karl Fischer an verschiedenen Lebensmitteln.
Deutsche Lebensmittel-Rundschau 77 (10) 343-355; 1981
3

2195 Zürcher, K.; Hadorn, H.
Störungen bei der Wasserbestimmung nach Karl Fischer.
Mitteilungen aus dem Gebiete der Lebensmitteluntersuchung und Hygiene 72 (2) 177-182; 1981
3

2196 Zuritz, C.; Singh, R.P.; Moini, S.M.; Henderson, S.M.
Desorption isotherm of rough rice from 10 °C to 40 °C.
Transactions of the American Society of Agricultural Engineers 22 (2) 433-436, 440; 1979
2,3,5

2197 Pixton, S.W.; Henderson, S.;
Moisture relations of dried peas, shelled almonds and lupins.
Journal of Stored Products, Research 15 (-) 59; 1979
5

2198   Pixton, S.W.; Warburton,S.
       The moisture content-equilibrium relative humidity
       relationship of rice bran at different temperatures.
       Journal of Stored Products, Research 11 (-) 1-8; 1975
       5
2199   Pixton, S.W.; Warburton,S.
       The relationship between moisture content and equilibrium
       relative humidity of dried figs.
       Journal of Stored Products, Research 12 (-) 87-92; 1976
       5
2200   Warburton, S.; Pixton, S.W.
       Moisture relations of freshly harvested barley in store.
       Journal of Stored Products, Research 9 (-) 269-272; 1973
       5
2201   Iglesias, H.A.; Chirife, J.
       Handbook of food isotherms: water sorption parameters for
       food and food components.
       Academic Press, New York; 1982
       1,2,5

3. Bibliography
Arranged according to subjects
The sub-chapters 3.1, 3.2, 3.3, 3.4 and 3.5 correspond to the code numbers 1, 2, 3, 4 and 5 always given in the last line of the cited references.

3.1. General descriptions (Code number 1)

6	7	10	12	13	14	52	53	77
85	87	117	150	164	212	215	218	224
236	237	278	281	287	288	291	312	339
356	368	375	383	408	439	444	477	484
489	490	493	494	524	525	529	532	534
535	536	538	573	581	583	598	602	605
623	633	638	660	661	662	663	667	669
728	745	749	758	771	794	815	843	844
845	846	847	848	849	874	881	886	892
894	918	922	927	936	944	950	952	971
990	991	994	998	1015	1016	1020	1022	1023
1027	1028	1030	1031	1040	1048	1049	1052	1073
1091	1110	1113	1114	1136	1137	1140	1142	1143
1144	1146	1157	1160	1172	1187	1199	1201	1202
1203	1208	1209	1210	1214	1215	1227	1243	1244
1250	1255	1256	1257	1258	1259	1260	1262	1263
1277	1278	1290	1308	1353	1354	1365	1371	1377
1406	1410	1413	1414	1415	1416	1418	1425	1454
1456	1472	1473	1507	1517	1529	1534	1578	1584
1585	1601	1615	1634	1641	1653	1676	1678	1679
1680	1684	1685	1686	1687	1688	1690	1691	1692
1694	1746	1747	1765	1766	1778	1782	1783	1789
1791	1806	1862	1863	1876	1877	1884	1890	1908
1910	1991	1997	1999	2017	2025	2032	2041	2042
2043	2045	2046	2047	2049	2068	2097	2102	2103
2106	2107	2146	2154	2172	2174	2201		

3.2. Thermodynamics of the sorption process (Code number 2)

Fundamentals, sorption models, hysteresis, mathematical description of sorption, sorption kinetics, diffusion and drying, water-binding phenomena

1	3	31	32	35	38	39	43	44
45	47	48	49	55	58	60	61	64
67	71	72	73	74	77	78	79	82
89	90	91	95	104	113	114	116	117
118	120	122	123	126	127	128	130	132
133	134	142	144	146	147	148	154	155
156	157	158	159	160	161	162	163	166
171	173	174	177	180	181	183	184	186
188	191	192	194	195	197	200	209	210
211	212	214	216	217	227	228	239	241
243	244	245	246	252	259	261	267	269
270	271	272	273	275	282	283	285	289
290	291	293	294	295	296	297	298	300
303	304	305	306	307	317	318	319	320
321	322	323	324	325	331	332	339	340
341	342	343	344	345	346	347	348	349
351	352	354	356	358	359	360	361	362
364	366	367	369	370	371	374	380	382
383	384	385	386	393	395	396	402	411
413	415	417	418	419	420	426	427	428
430	437	443	452	453	455	461	466	470
471	472	474	477	479	480	481	482	484
485	486	487	492	493	494	497	498	502
508	509	511	512	513	514	516	518	524
525	529	531	532	533	534	535	541	544
545	546	548	549	551	552	555	558	559
567	568	569	570	572	573	574	576	579
581	586	587	588	590	591	594	595	597
598	600	601	607	608	611	612	621	624
625	626	627	630	632	633	635	636	637
638	647	648	649	650	651	654	655	656
657	658	660	665	666	670	671	673	674
683	684	685	689	693	695	697	698	699
706	708	709	711	713	715	717	726	727
737	738	739	740	743	749	758	759	761
763	765	766	768	769	770	771	773	774
776	777	788	790	791	794	795	796	798
802	805	807	808	811	814	820	824	825
826	835	840	843	851	852	853	857	859
860	861	865	866	867	868	869	883	884
885	886	887	888	889	890	891	895	896
897	898	904	921	923	924	925	932	935
936	937	938	939	940	941	942	943	944
947	948	949	950	951	953	954	955	956
970	980	982	983	987	990	991	992	997
1000	1004	1005	1010	1011	1012	1016	1019	1020
1022	1023	1030	1033	1038	1039	1042	1047	1050
1052	1062	1063	1065	1067	1070	1073	1080	1081
1082	1086	1090	1093	1097	1098	1101	1106	1110
1115	1116	1118	1119	1120	1121	1122	1123	1126
1127	1128	1129	1130	1136	1143	1145	1148	1154
1157	1168	1169	1172	1174	1175	1176	1177	1178
1179	1181	1183	1184	1185	1187	1191	1192	1193
1216	1218	1220	1221	1223	1224	1229	1232	1241
1242	1246	1248	1255	1258	1264	1276	1284	1285
1286	1287	1288	1290	1292	1293	1294	1295	1303
1309	1316	1324	1325	1332	1333	1337	1340	1341
1343	1345	1349	1353	1358	1360	1366	1367	1369
1371	1381	1382	1384	1388	1391	1393	1394	1403
1404	1406	1407	1408	1409	1411	1414	1417	1421
1429	1430	1431	1432	1433	1437	1439	1441	1442

```
1443  1446  1447  1448  1449  1450  1455  1456  1457
1460  1461  1462  1463  1465  1474  1475  1477  1478
1495  1498  1500  1501  1508  1510  1511  1512  1513
1514  1515  1516  1517  1520  1521  1522  1523  1524
1525  1527  1533  1534  1535  1536  1538  1539  1540
1545  1546  1550  1551  1553  1554  1555  1556  1557
1560  1564  1567  1591  1599  1600  1606  1611  1619
1620  1628  1629  1630  1632  1633  1639  1646  1647
1649  1652  1653  1654  1672  1676  1678  1679  1680
1682  1683  1705  1706  1711  1712  1713  1714  1716
1717  1718  1719  1722  1727  1730  1731  1732  1733
1734  1735  1736  1737  1739  1740  1743  1746  1751
1752  1756  1757  1758  1771  1772  1775  1776  1777
1782  1783  1784  1786  1787  1789  1791  1795  1796
1798  1801  1807  1808  1809  1819  1822  1823  1824
1825  1831  1832  1833  1834  1835  1837  1843  1846
1848  1855  1856  1857  1858  1859  1860  1861  1862
1869  1872  1874  1879  1881  1884  1885  1890  1891
1893  1894  1895  1903  1910  1914  1915  1917  1918
1919  1920  1921  1926  1931  1933  1934  1938  1939
1940  1942  1945  1946  1947  1950  1955  1958  1959
1960  1966  1967  1973  1977  1985  1987  1991  1994
1996  1997  1999  2001  2002  2004  2006  2017  2020
2023  2026  2028  2030  2036  2039  2040  2042  2043
2044  2045  2046  2047  2049  2055  2057  2063  2064
2070  2071  2072  2073  2076  2079  2080  2083  2084
2085  2094  2097  2098  2100  2101  2104  2110  2112
2121  2123  2125  2126  2137  2141  2150  2152  2153
2156  2167  2170  2172  2174  2175  2176  2177  2182
2183  2184  2185  2186  2188  2189  2191  2196  2201
```

3.3 Measuring methods (Code number 3)

Fundamentals, equipment, influencing factors (adjustment of relative humidity, determinations of the water content, etc.)

2	5	30	34	37	39	45	46	50
68	77	78	81	85	86	87	91	92
93	96	97	98	106	109	112	116	117
129	130	136	137	139	152	157	158	159
169	178	179	182	189	196	205	207	208
212	213	218	220	227	228	229	230	232
233	235	240	248	253	254	255	258	260
264	265	266	268	269	270	274	276	283
286	292	301	303	305	309	311	314	316
317	318	319	320	333	334	358	363	364
369	371	372	377	392	399	402	404	415
430	434	438	442	444	451	452	454	458
470	475	477	478	483	486	490	491	493
495	500	511	519	527	528	530	534	544
556	557	573	575	582	588	592	593	596
603	611	612	614	623	626	639	640	655
656	659	660	664	667	668	685	687	688
692	693	697	698	702	703	704	705	707
709	711	714	718	719	720	721	731	732
733	734	735	736	745	746	747	748	752
754	771	772	773	776	777	780	784	812
813	814	817	818	819	822	827	829	830
835	836	840	857	861	865	871	878	879
882	888	889	893	894	897	903	909	910
917	918	919	921	926	927	929	934	935
959	961	963	964	968	972	976	979	984
985	986	987	993	997	1005	1010	1023	1024
1026	1031	1033	1036	1041	1047	1050	1055	1058
1059	1061	1063	1064	1066	1067	1068	1069	1070
1074	1083	1084	1095	1102	1107	1110	1113	1132
1133	1134	1147	1153	1154	1170	1171	1173	1180
1195	1208	1210	1212	1213	1217	1219	1226	1232
1233	1236	1239	1254	1262	1267	1282	1283	1289
1294	1295	1296	1298	1299	1301	1302	1308	1309
1311	1312	1313	1316	1324	1329	1332	1334	1336
1337	1340	1346	1347	1350	1353	1357	1370	1372
1373	1374	1376	1380	1387	1399	1408	1410	1412
1415	1418	1419	1420	1421	1423	1424	1425	1426
1441	1449	1451	1452	1453	1454	1456	1458	1465
1467	1468	1469	1471	1476	1485	1486	1491	1494
1496	1497	1503	1505	1508	1526	1528	1529	1530
1534	1547	1549	1571	1578	1592	1601	1602	1610
1630	1636	1637	1638	1648	1651	1661	1662	1663
1664	1667	1668	1669	1670	1671	1672	1676	1677
1679	1680	1685	1688	1691	1692	1694	1695	1698
1700	1701	1702	1714	1715	1720	1721	1749	1750
1751	1759	1760	1762	1764	1766	1771	1772	1774
1779	1780	1785	1787	1793	1799	1804	1810	1821
1826	1830	1833	1834	1835	1836	1838	1841	1843
1845	1853	1865	1878	1880	1881	1883	1884	1892
1897	1898	1899	1901	1904	1905	1909	1910	1911
1917	1926	1927	1928	1929	1931	1948	1949	1957
1959	1960	1965	1969	1971	1972	1974	1979	1981
1988	1989	1993	2003	2011	2015	2016	2017	2022
2027	2030	2037	2046	2051	2053	2054	2055	2056
2060	2062	2065	2069	2074	2077	2079	2083	2090
2096	2099	2103	2105	2106	2108	2109	2111	2114
2115	2116	2117	2118	2119	2120	2128	2130	2131
2133	2137	2138	2139	2142	2143	2144	2145	2147
2148	2149	2151	2155	2156	2157	2160	2161	2165
2168	2169	2173	2174	2178	2181	2192	2194	2195
2196								

3.4 Influence of water activity on product stability (physical, chemical, microbiological) (Code number 4)

4	6	8	9	10	11	12	13	14
15	16	17	18	19	20	21	22	23
24	25	26	27	28	29	33	40	42
48	52	53	54	56	57	59	62	63
67	69	75	76	82	83	84	85	87
88	94	97	101	102	108	110	111	115
116	117	121	124	125	131	135	138	140
141	143	153	156	165	168	172	185	187
190	199	201	202	203	204	205	206	217
219	221	225	226	231	234	237	238	239
242	249	250	251	252	256	257	261	262
263	277	278	279	280	281	284	293	298
301	302	304	310	313	321	324	326	327
328	332	335	336	337	350	354	355	356
357	365	367	369	371	373	374	375	376
377	379	381	386	387	388	389	391	394
395	396	397	398	399	400	401	402	403
404	405	406	407	408	409	410	411	412
414	421	422	423	424	425	429	431	435
436	437	442	445	446	447	448	449	450
451	456	457	459	462	463	465	468	469
477	486	489	495	496	499	501	503	504
505	515	520	521	522	523	526	528	533
534	537	539	540	542	543	547	550	553
560	561	562	563	564	565	566	571	577
578	580	583	584	585	593	594	595	599
601	604	608	609	610	613	615	616	618
620	623	628	629	631	641	642	643	645
646	648	649	652	653	679	680	681	682
683	686	691	694	696	700	701	706	710
712	716	729	735	741	750	751	753	754
755	756	757	760	762	764	767	771	772
778	779	781	782	783	758	786	787	789
792	801	802	803	804	805	806	809	810
811	812	814	815	816	821	823	828	831
832	833	834	837	838	841	842	843	844
845	846	847	848	849	850	854	855	856
858	859	866	859	870	872	873	875	876
877	879	884	899	900	902	906	907	911
912	913	914	915	916	920	927	928	930
931	933	957	958	965	966	967	969	973
974	975	978	979	981	988	989	996	1001
1002	1003	1005	1006	1007	1008	1013	1014	1015
1016	1017	1018	1020	1021	1022	1023	1024	1025
1028	1029	1031	1032	1034	1035	1036	1037	1040
1042	1056	1057	1060	1071	1072	1075	1076	1077
1079	1080	1087	1088	1089	1092	1096	1098	1099
1105	1106	1108	1112	1124	1125	1126	1131	1135
1137	1138	1139	1141	1144	1146	1149	1150	1151
1152	1155	1156	1158	1159	1160	1161	1162	1163
1164	1166	1167	1172	1174	1182	1186	1187	1188
1189	1190	1191	1194	1196	1197	1200	1203	1204
1205	1206	1207	1208	1209	1210	1211	1214	1215
1216	1222	1225	1230	1238	1240	1246	1251	1252
1253	1254	1255	1257	1259	1260	1261	1264	1265
1268	1269	1270	1271	1272	1273	1274	1277	1278
1279	1280	1281	1291	1300	1304	1305	1306	1310
1313	1314	1315	1316	1317	1318	1326	1327	1330

```
1331  1337  1339  1348  1354  1355  1364  1368  1375
1378  1379  1382  1383  1385  1389  1390  1392  1395
1396  1397  1398  1399  1400  1401  1402  1405  1413
1414  1416  1422  1428  1435  1436  1440  1444  1445
1456  1464  1470  1479  1480  1481  1482  1483  1484
1487  1488  1489  1493  1506  1509  1410  1511  1515
1516  1517  1518  1519  1525  1531  1537  1538  1539
1541  1542  1543  1544  1548  1561  1562  1576  1577
1580  1581  1582  1583  1584  1585  1586  1587  1588
1589  1590  1591  1592  1595  1596  1597  1598  1603
1607  1608  1609  1612  1613  1614  1617  1618  1621
1622  1623  1624  1626  1629  1634  1637  1638  1640
1658  1659  1665  1666  1673  1680  1681  1684  1686
1687  1688  1689  1690  1693  1696  1697  1699  1701
1703  1704  1708  1709  1710  1723  1724  1726  1728
1729  1738  1739  1740  1741  1743  1744  1745  1747
1748  1761  1763  1767  1768  1769  1770  1781  1788
1790  1792  1794  1797  1800  1802  1805  1806  1807
1811  1812  1813  1814  1815  1816  1817  1818  1819
1820  1827  1828  1852  1854  1864  1866  1867  1868
1869  1871  1872  1873  1882  1886  1887  1888  1889
1890  1896  1906  1907  1912  1913  1922  1923  1924
1925  1932  1937  1938  1941  1942  1943  1951  1952
1953  1956  1962  1963  1968  1976  1978  1982  1983
1984  1986  1989  1990  1995  1997  1998  2000  2005
2007  2008  2009  2010  2012  2013  2014  2017  2018
2019  2025  2026  2031  2032  2033  2034  2041  2048
2052  2057  2058  2059  2060  2061  2065  2067  2068
2075  2078  2081  2082  2087  2091  2092  2093  2095
2100  2101  2113  2124  2126  2127  2129  2132  2134
2136  2140  2154  2164  2170  2171  2179  2180  2187
2190  2193
```

3.5 Sorption data (Code number 5)

3.5.1 Sorption data arranged according to the consecutive numbers of the references

2	4	6	9	10	12	13	14	16
17	19	21	22	28	34	35	36	38
39	41	43	45	47	49	51	55	57
59	62	65	66	70	76	78	80	81
83	84	86	88	89	90	91	99	100
103	104	105	107	108	109	119	121	123
131	132	133	134	139	143	144	145	147
149	151	152	160	166	167	169	170	171
173	174	175	176	177	181	188	190	193
195	196	198	200	205	210	212	222	223
225	227	233	240	247	248	249	252	260
261	266	269	273	276	277	285	293	294
299	301	304	305	306	308	313	315	324
325	329	330	331	338	342	353	355	359
360	361	362	366	367	373	374	376	378
380	382	387	390	393	394	395	396	398
402	413	415	416	417	418	419	420	424
426	429	431	432	433	438	440	441	442
445	446	450	451	453	460	461	463	464
467	471	472	473	474	476	477	479	482
486	487	488	489	490	493	495	500	501
506	507	508	510	515	517	518	519	521
528	532	533	534	537	540	543	548	549
552	554	568	569	570	573	574	578	584
589	590	592	599	601	602	603	604	605
606	607	614	615	617	619	621	622	626
627	630	633	634	644	651	652	655	657
658	659	660	661	670	672	673	675	676
677	678	681	683	689	690	691	693	696
702	707	711	713	719	720	722	723	724
725	730	739	740	742	743	744	746	751
753	757	760	763	766	769	770	771	773
775	776	777	781	784	789	792	793	794
796	797	799	800	801	807	808	809	811
812	813	814	815	816	820	824	826	827
838	839	841	842	844	846	850	851	852
853	855	856	857	860	861	862	863	864
874	875	876	880	881	882	883	884	889
890	891	892	893	894	897	898	901	904
905	908	909	910	916	917	920	921	922
923	924	925	927	929	932	935	936	938
941	942	945	946	947	950	951	952	953
954	956	960	962	970	976	977	978	981
983	990	994	995	997	999	1001	1002	1003
1004	1005	1009	1010	1011	1014	1015	1016	1033
1036	1038	1043	1044	1045	1046	1047	1051	1053
1054	1056	1057	1061	1063	1064	1068	1076	1078
1079	1080	1083	1085	1086	1093	1094	1100	1103
1104	1105	1107	1108	1109	1111	1114	1115	1117
1118	1119	1123	1128	1130	1133	1156	1157	1161
1164	1165	1169	1170	1171	1174	1175	1176	1180
1182	1183	1187	1190	1191	1193	1195	1198	1202
1211	1223	1228	1229	1230	1231	1234	1235	1236
1237	1241	1244	1245	1247	1249	1250	1254	1255
1257	1260	1262	1263	1264	1266	1274	1275	1276
1284	1286	1289	1291	1293	1295	1296	1297	1298
1303	1307	1308	1309	1310	1315	1316	1318	1319
1320	1321	1322	1323	1328	1330	1332	1334	1335
1336	1338	1340	1341	1342	1343	1344	1347	1348
1351	1352	1353	1356	1358	1359	1361	1362	1363
1364	1365	1366	1368	1371	1374	1380	1381	1383
1384	1385	1386	1388	1394	1395	1401	1402	1406

1414	1415	1416	1417	1420	1425	1426	1427	1430
1434	1437	1438	1441	1442	1443	1444	1449	1450
1451	1452	1453	1454	1455	1456	1457	1458	1459
1460	1463	1464	1465	1466	1474	1476	1477	1478
1479	1484	1485	1486	1488	1489	1490	1491	1492
1497	1499	1500	1502	1504	1506	1509	1511	1518
1521	1522	1523	1524	1526	1530	1532	1534	1536
1538	1543	1544	1546	1549	1552	1558	1559	1563
1565	1566	1567	1568	1569	1570	1571	1572	1573
1574	1575	1577	1578	1579	1580	1583	1584	1593
1594	1599	1604	1605	1606	1607	1610	1611	1614
1616	1617	1619	1621	1625	1627	1628	1629	1630
1631	1632	1633	1635	1636	1637	1638	1642	1644
1647	1652	1653	1657	1660	1661	1666	1674	1675
1676	1678	1680	1682	1683	1690	1693	1697	1699
1705	1706	1707	1714	1715	1717	1718	1719	1723
1724	1725	1727	1731	1732	1733	1735	1739	1740
1741	1742	1744	1745	1746	1753	1754	1755	1756
1757	1758	1764	1765	1770	1771	1773	1774	1776
1778	1784	1786	1797	1800	1801	1803	1807	1808
1809	1811	1822	1823	1824	1829	1839	1840	1842
1844	1845	1847	1848	1849	1850	1851	1853	1854
1857	1865	1867	1869	1870	1871	1872	1874	1875
1876	1877	1880	1881	1883	1885	1888	1890	1891
1894	1900	1902	1903	1909	1910	1911	1912	1915
1916	1924	1926	1929	1930	1931	1933	1934	1935
1936	1939	1941	1944	1950	1953	1954	1955	1958
1959	1961	1962	1963	1964	1970	1972	1974	1975
1977	1980	1982	1984	1985	1992	1996	1997	1999
2004	2019	2021	2022	2024	2026	2029	2030	2035
2038	2039	2043	2046	2047	2050	2053	2054	2055
2058	2062	2065	2066	2069	2070	2072	2073	2074
2076	2077	2079	2084	2085	2086	2087	2088	2089
2097	2098	2103	2106	2108	2110	2113	2122	2127
2128	2132	2135	2139	2141	2155	2156	2157	2158
2159	2160	2161	2162	2163	2166	2174	2175	2176
2177	2179	2189	2193	2196	2198	2199	2200	2201

3.5.2 Sorption data arranged alphabetically according to products

Please note: numbers marked by * refer to a single water activity-value only (not to a complete sorption isotherm) of the product concerned

- juice	378*	681	844	1463	2201			
- pectin	1463							
Apricot	378*	617	681	696	816	844	1348	1765
	1800	1984	1985	2201				
- kernel	617							
- marrow	1765							
- milk-beverage	569							
Arabate, calcium	1629							
Arachidonic oil	1254	1262						
Areca nut	1845							
Arginine	55							
Arrow root, starch	1739							
Artichoke	378*							
Ascorbic acid-lactose (mixture)	1190	1191						
- mannitol-starch (mixture)	1190	1191						
Asparagine	55							
Asparagus	378*	681	844	1463	1765	2201		
Aubergine	2158	2159	2201					
Avocado	378*	1248	2158	2159	2201			
Bacillus stearothermophilus, spores	2098							
Bacillus subtilis, spores	472							
Bacon	152*	844	1104					
Baker's yeast	838	844	1231	1463	1844			
Baking powder	844							
Banana	378*	528	681	691*	794	838	844	947
	1463	1849	1964	2158	2159	2201		
- flour	1964							
- IMF	976*							

Carrageenan	2085						
Carrot	76	378*	569	617	678	681	691*	696
	713	811	844	976*	1230	1231	1234	1247
	1250	1309	1452	1453	1463	1676	1745	1926
	1959	2201						
- IMF	976*						
- seed	678	2201					
Casein	109	166	169	170	171	173	614	670
	784	904	929	1223	1245	1254	1260	1262
	1359	1606	1628	1632	1717	1723	1727	1731
	1851	1999	2201					
Caseinate	2	1085*	2110*	2201			
Casein-glucose (mixture)	1182						
Cashew nut	200	724	1074*	1504	1614	1964	2201
Cassava	1890						
- starch	2139						
Cauliflower	378*	569	844	892	893	894	
Celery	378*	691*	693	838	844	1770	2158	2159
	2201							
Cellobiose	1414						
Cellophane	1885	2135					
Cellulose	149	210	450	532	655	661	740	813
	814	979	1038	1223	1289	1394	1416	1455
	1511	1717	1840	1926	2029	2038	2039	2072
	2074	2079	2157	2160	2161			
- acetate	149	824	2039	2139			
- carotene (mixture)	108						
- glucose-glucose oxidase (mixture)	19						
- microcrystalline	109	227	898	951	960	1430	1911
	2074	2157	2160	2161	2201		
- oleoresin	1014						
Cereals	212	1244					
Chanterelle	2158	2159					
Cheese	107*	225	301*	578	754	815*	844	881*
	1074*	1318*	1322*	1323*	1334	1335	1484*	1602*
	1724	1726*	1929*	2016*	2060*	2158	2159	2201
- Appenzell	1318*	1724*					

- Beaumont   1318*

- Belle de champs   1724*

- Bel paese   1724*

- Bergkäse   1724*

- Bola   1318*

- Brie   1318* 1724*

- Camembert   815* 1318* 1724*

- Cheddar   1318* 1602* 1724*

- Cottage   1724*

- Danablau   1318* 1952*

- Edam   1318* 1724* 2158 2159 2201

- Emmental   815* 1318* 1335 1724 2158 2159 2201

- Fontal   1724*

- Fontina   1318*

- Gouda   815* 1318* 1724*

- Gruyere   1318* 1334 1724*

- Limburg   1724*

- Mimolette   1318*

- Mozzarella   1318*

- Münster   1318* 1724*

- Parma   1318* 1724*

- Pirinees   1318*

- Port salut   1318*

- powder   107

- Provolone   1318*

- Raclette   1318*

- Roquefort   1318*

- Rotschmierkäse   1724*

- Saint-Abray   1318*

- Saint-Paulin 1724*

- Saint-Varent 1318*

- sandwich   1016

Coconut   1053 1964

- coarse meal   1972

- milk   789

- oil   1607

Cod   359   360   378* 440   441   460   506   508
                        605   811   844   970   1453 1463 1959 2201

- skin   441

Coffee   86   107   313   451   476   489   568   681
                        720   725   757   826   844   1117 1170 1231
                        1263 1266 1406 1453 1463 1605 1614 1765
                        1963* 1964 2004 2132 2163 2201

- bean   451   476   720   725   1406 2132 2201

- extract   1765 2201

- instant   1170 1963*

- surrogate   844   1231

Collagen   294   426   471   472   1011 1123 1241 1276
                        1284 1953 1954 2201

Condiments   476   693   725   2158 2159 2192* 2201

                        see also anise, cardamom, cinnamon, clove,
                        coriander, garlic, ginger, laurel, marjoram,
                        nutmeg, pepper, pimento, savory, thyme

Confectionary   315   815   844   850   1005 1274 1484*

Cookie   2179

Copra   308   724   927   1406 1569 1707 2201

Coriander   2158 2159 2201

Corn   see maize

Corned beef   152* 500* 578   1026*

Corn flakes   1015

Cotton   89   2030

- seed   1043 1044 1057 1381 2201

Cranberry   378*

Cream   378*

Cress, seed   1463

Crispbread   844   1005 1076 1453

Cucumber   378*

Curd   103   104   844   1453 2201

- syrup- saccharose	1974 (mixture)						
Glue	742						
Glutamate	55	1954	2201				
Glutamic acid	55	1675					
Glutamine	55						
Gluten	305	418	419	420	775	1463	2201
Glycerol	424 1871	489 1874	521	742	1476	1485	1486	1869
Glycine	651	1051	1824				
- polymers	1064	1358	1361				
Glycogen	1478						
Gooseberry	378*						
Grain	923	1742					
Grape	378*	1800					
- fruit	378*	938	2158	2159	2201		
- juice	378*						
Groats	1742						
Guar gum	2085	2109					
Guava	1464	1849					
- corn flakes (mixture)	107						
- taro (mixture)	1464	2201					
Gum arabic	1629	2109*					
Halibut	100	2201					
Ham	152*	394	578*	815*	1026*	1114*	1115*	1202* 1484*	1485*	1486*	1523*	1695*	1699*	2051*	2154*
- raw	152*	1202					
Hazelnut	86	781	844	1617	2201		
Hemoglobine	266	331	510	570	1953		
Hibiscus	2158	2159	2201				
Histidine	55						
Honey	247 2060*	301*	517	578*	844	1328	1484*	1725

Hop	519	862	1463	1636	2201		
Horse radish	693	1341	2158	2159	2201		
Humic acid	1536						
Hyaluronic acid	223						
Insulin	1954						
Invert sugar	489	568	844	1386	1476		
Isoleucine	55						
Jam	301*	500*	578*	815*	963*	1399*	1484*	2060*
Jelly	316						
Juice, dried	1963*						
Kale	713						
Karaya	2085						
Kelp	151						
Keratin	784	2097	2098				
Ketchup	2060*						
Lactose	81	169	170	174	176	568	627	630
	844	1245	1254	1257	1260	1262	1263	1453
	1459	1460	1869	1870	1871	1894	1926	2088
	2089	2103	2128*	2201				
Lard	908						
Laurel	673	938	941	2158	2159	2201	
Leek	378*	713	844	1228	1234	1371	1452	1453
	2201							
Lemon	378*	1105	1996	2201			
Lentil	1463	2158	2159	2201			
Lettuce	378*						
Leucine	55	261	651				
Levulose	568	1386	1453	1894			
Lignin	983						
Lima bean	378*						
Linseed	678	927	1057	1463	1569	1888	2201
Lipids	1511						
Lipoprotein, serum	510						

Locust bean	2085							
Lupin	2197	2201						
Lysine	55	1674						
Lysozyme	240	890	891	1193				
Macadamia nut	724	809	2162					
Mace	693							
Mackarel	962							
Magou	107							
Maize	100	188	276	361	362	417	418	419
	420	432	493	619	672	720	722	723
	776	777	792	794	860	917	922	927
	932	1128	1371	1406	1415	1454	1456	1463
	1485*	1497	1534	1563	1568	1682	1683	1742
	1808	1839	1848	1916	1933	1934	1980	2109*
	2110*	2174	2176	2201				
- amylose	1394							
- endosperm	418	419	420	689	1119	1839	1847	
- feed	1578							
- flakes	1015							
- meal	84	107	844	889	1118	1484	2054	2201
- starch	107	418	419	420	856	857	921	960
	994	1009	1175	1176	1223	1332	1420	1426
	1427	1451	1453	1463	1739	1740	1773	1774
	1883	1958	2024	2053	2201			
- starch hydrolysate	1427							
- syrup	424	568	1869	1870	1871			
Malt	1047	1237	1356	1453	1566	2022		
- coarse meal	6	9	10	12	21	22		
- coarse meal-lecithine (mixture)	6	9	10	11	22			
- coarse meal-monophenylphosphate (mixture)	21							
Maltodextrin	1366	1977						
Maltopentose	1466							
Maltose	81	521	627	630	839	844	1414	1416
	1466	1894	2103	2106	2201			
Maltotriose	521	1414						

Mango	378*	976	1849				
- chutney	2060*						
- custard	1849						
- IMF	976*						
Manioc, starch	2043						
Mannit	2108						
Maracuja	324						
Marcipan	533	844					
Margarine	2016*						
Marjoram	693	2158	2159	2201			
Marrow	681	2201					
Marshmallow	315						
Mate	1964						
Matzon	107						

Meat   301* 450   500* 844   964* 1115* 1198* 1202*
              1231 1463 1484* 1583 1584 1602* 1692* 1695*
              1764 1765 1964 2076* 2077* 2155*
       see also bacon, beef, brawn, chicken, ham, mutton,
       pork, turkey, veal

Melon	378*						
Methionine	55						

Milk   107   169   170   174   277   378* 533   696
              792   815* 844   852   853 1074* 1156 1245
             1453 1484* 1753 1974 2109* 2110* 2201

- babyfood	2055	2201					
- orange-drink	177	2201					

- skimmed   6   107   169   170   174   533   796   838
              844   863   875   876 1228 1230 1231 1237
             1245 1453 1463 1563 1944 2088 2201

Millet	723	1463	1742				
Miso	796	800	2170*				
Molasses	382*						
Mucine	193						
Muco polysaccharide	195						
Mullet	353						

```
- powder-            702
  starch syrup (mixture)

Orgeat           1559  2201

Ornithine          55

Palatinite       2108

Palmkernel        724  1406  1900

Papaya            325   378*  486   487  1464  2201

Paprika           693   725   844  1014

Parsley           844  1228  1231  1237  2201

Parsnip           200   222   378*  495   678  1463  2201

Pasta               6    22   107   205   696   838   844   875
                  876  1231  1463  1485  1486  1572  1842  2139
                 2158  2159  2201

- macaroni        696   844  1463  1572  2139  2201

- spaghetti       107

Pastry, puff      633

Pea               277   378*  394   476   453   476   543   569
                  678   681   724   811   838   844  1165  1237
                 1452  1453  1742  1745  1770  1959  2197  2201

- flour           107  1569  2201

- seed            678

Peach             378*  533   615   691*  696   743   844   860
                  938  1745  1758  1800  2201

Peanut             86   139   479   619   724   844   884   927
                 1046  1406  1463  1485  1486  1563  1569  1865
                 2175  2201

- flakes          884

- oil            1260  1263  2201

- pod              35  2175  2201

Pear              378*  691*  947  2158  2159  2201

Pecan nut         200   724  2158  2159  2201

Pectate           196   198

Pectic acid       196  1521

Pectin            109   196   450   844  1211  1223  1237  1463
                 1518  1754  1926  2085  2110*
```

Pepper	86	476	691*	693	725	844	1005	1485*
	1486*	2127	2166					
- green	378*	615	1744	1745	2201			
- seed	617							
Peppermint, herb	2158	2159	2201					
Persimmon	378*	617						
Phenylalanine	55							
Pimento	86	476	725	1453	2201			
Pineapple	378*	844	947	2158	2159	2201		
Pineapple, juice	1488							
Polyethylene-glycol	1869	1870	1871					
Polymeres, natural	1053	1054						
- synthetic	119							
Polyvinyl-chloride	2072							
Popcorn	1056							
Pork	153*	378*	676	811	844	1463	1765	1959
	2156	2201						
- cured smoked	153*	1765						
- glycerol (mixture)	301	1577						
Potassium chloride	1869	1870	1871					
Potato	378*	453	533	540	568	615	681	696
	711	713	746	811	815*	844	901	936
	981	1231	1247	1308	1309	1340	1342	1371
	1463	1491	1653	1676	1745	1755	1757	1758
	1765	1926	1935	1959	2201			
- chips	617	844	1056					
- fibre	901							
- flakes	815*	1231	1935	2201				
- flour	1765							
- juice	901							
- mashed	1491							
- protein	901							

```
- starch          109   293   366   533   537   592   622   844
                  857   892   894   901   960  1183  1316  1394
                 1414  1437  1438  1451  1453  1463  1465  1619
                 1739  1773  1774  1776  1847  1926  2043  2046
                 2047  2110* 2185  2160  2201

- starch gel     1926

Proline            55

Promine           874  1237

Propylene        1869  1870  1871
glycol

Protein
       see also actomyosin, albumen, bean leaf protein, bean
       protein isolate, casein, collagen, edistin, elastin,
       fish protein concentrate, fish protein isolate, gela-
       tine, globuline, keratin,lipoprotein serum, myosin,
       potato protein, promine, protein isolate, safflower
       protein, salmine, sarcoplasma, single cell protein,
       soybean protein, tropocollagen, whey protein

- isolate        1912

Prune             233   378*  617   681   696   707   860  1228
                 1371  1463  1800  1984  2201

Pudding powder    107   614

Pumpkin          378*  1463

Radish           378*  1463  2158  2159  2201

Raffinose        1414

Raisin            233   681   696   844   860  2016* 2201

Rape             1549

- seed            724  1463  1563  1565  1575  2201

Raspberry          45   378*  569   681  1463  1926  2201

- juice          378*  681

Rayon            2029

Rhubarb          378*  681  1463  2201

Ribonuclease     1193

Rice               34    35   105   269   277   432   564   606
                  619   722   723   792   794   811   815*  844
                  861   897   909   910   999  1003  1033  1045
                 1371  1380  1406  1453  1463  1610  1611  1627
                 1630  1742  1902  1929* 1939  1959  1963* 2021
                 2156  2196  2201

- bran            626  1045  1563  2198  2201

- flour          1963*
```

- groats	1902						
- Paddy	1610						
- starch	1438	1773	1774				
Roe	353	916*					
Rusk	107	633*	842	844	1365	1399*	1453	1463
- cracker	696	1056	1170	1484*			
Rye	432	461	1237	1463			
- bran	1778						
- flour	844	1076	1451	1453	1985		

Saccharose
```
212    489    507    568    839    844   1254   1257
1260   1262   1263   1310   1338   1386  1476   1511
1854   1869   1870   1871   1874   1880  1890   1894
1926   1936   1964   2106   2128*  2201
```

Saccharomyces   1546   2201
cerevisiae

Safflower-   950   2201
protein

Sago	1875	2158	2159	
Salmine	294	1676	2201	
Salmon	844	1161	1363	2201
Sarcoplasma	1526	2201		
Sauerkraut	838	844	1247	1770

Sausage
```
143*   152*   153*   431*   490*   500*   578*   935
963*  1026*  1074*  1114*  1115*  1202*  1399*  1425*
1485* 1486*  1602*  1695*  1699*  1760*  1952*  2053*
2060* 2074*
```

- Bierwurst	152*				
- Blutwurst	500*	1026*	1115*	1202*	
- Bologna-type	935	1202*	1699*		
- Bratwurst	1026*				
- Cervelat	152*	1026*			
- dry	490*	1114*	1399*	1699*	2053*
- Fleischwurst	152*	1114*			
- Gelbwurst	1115*				
- Jägerfleisch	1114*				
- Jagdwurst	1114*				

- Landjäger   578*

- Liver   153*   500*   578* 1026* 1114* 1115* 1202*

- Liver, pate   153*

- Luncheon meat 153* 1602*

- Lyoner Wurst 1115*

- Meat loaf   153* 1115* 1695*

- Mettwurst   152* 1115*

- Mortadella   152* 1115*

- Paprika Wurst   1115*

- Plockwurst   143*

- Salami   143*   152*   500*   578*   963* 1026* 1074*
                       1115* 2074*

- Teewurst   1115*

Savory   693   2158   2159   2201

Semolina   205   1451   1453   1599   1742

Serine   55

Sesame, seeds   416   1463

Shrimp   277   464

Silicagel   59   212   990

Single cell   1574
protein

Sodium   1223
alginate

Sodium   841
bitartrate

Sodium   841   1308
carbonate

Sodium   2   614   874   2155
caseinate

Sodium   78   661   962   1765   1869   1870   1871
chloride

Sodium   398
citrate

Sodium   398
ethyldiaminacetate

Sodium   1231
glutamine

Tartrate, calcium	390							
Tartrate, sodium	841							
Tea	681	720	725	844	977	978	1156	1453
	1463	1964	2158	2159	2201			
Teff (eragrostis tef)	722	1502						
Threonine	55							
Thyme	938	941	2158	2159	2201			
Tomato	39	249	338	378*	681	691*	838	844
	875	876	889	963*	1005	1228	1237	1275
	1383	1452	1453	1463	1770	1926	1964	2201
- concentrate	378*	889	963*	1228	1237			
- ketchup	963*							
Toprina	1574	2201						
Tortilla IMF	578	1543						
Tragacanth	1847	2110*						
Triticale	41*	723						
Tropocollagen	1284	1286						
Trout	2158	2159	2201					
Tryptophane	55							
Tuna	1100	2201						
Turkey	1082	1083	2201					
Turnip	378*	1463						
Tyrosine	55							
Urease	1867							
Valine	55							
Veal	892	894						

Vegetable

see artichoke, asparagus, aubergine, bean, beet
root, black salsify, broccoli, brussels sprouts,
cabbage, carrot, cassava, cauliflower, celery,
chickpea, chicory, chive, cucumber, endive, fennel,
horse radish, leek, lentil, lettuce, marrow, onion,
paprika, parsley, pea, pepper green, potato, pump-
kin, radish, rhubarb, sauerkraut, soybean, spinach,
swedish turnip, sweet potato, tomato, turnip

Wafer	121	633*	842	2201			

Walnut	200	844	951	1263	1463	1676	1929* 2016*
	2035	2058	2201				

Watermelon	378*

Wheat	86	90	91	145	147	190	277	299
	304	418	419	420	432	438	450	461
	482	589	606	619	673	675	690	722
	775	792	794	820	860	905	917	927
	1180	1237	1336	1371	1399*	1414	1417	1443
	1463	1509	1530	1549	1563	1565	1567	1568
	1571	1579	1666	1877	1950	1980	2026	2109*
	2110*	2174	2177	2176	2201			

- bran	1778	1972

- feed	2086	2201

- flour	6	22	66	91	107	305	306	467
	473	589	771	775	838	844	875	876
	951	1169	1187	1336	1388	1399*	1451	1453
	1492	1599	1822	2201				

- gluten	775

- middling- dried yolk (mixture)	6	22

- starch	109	248	305	306	775	844	1425*	1451
	1453	1619	1773	1774	1847	1961	2043	2047
	2185	2201						

Whey	169	170	844	1079	1236	1237	1245	1453
	1463	1741	1829	2074*	2109*	2110*	2201	

- cheese	167	1245

- protein	166	181	739	874	2201

- rennet	844

- soy beverage	177	2201

- sweet	169	1741	2201

Wood	89	1063

Xanthane	2085

Xylitol	51	1009	1111	1453	2066	2108

Yam	83	696	1309	1926	2201

Yeast	6	681	1478	1546	1574	2201

Yellow boletus	838	844	875	876	1231

Yogurt	1319*	2158	2159	2201

Yolk, egg	677	1463	1853